Marcilio C.P. de Souto   Maricel G. Kann (Eds.)

Advances in Bioinformatics and Computational Biology

7th Brazilian Symposium on Bioinformatics, BSB 2012
Campo Grande, Brazil, August 15-17, 2012
Proceedings

 Springer

Series Editors

Sorin Istrail, Brown University, Providence, RI, USA
Pavel Pevzner, University of California, San Diego, CA, USA
Michael Waterman, University of Southern California, Los Angeles, CA, USA

Volume Editors

Marcilio C.P. de Souto
Universidade Federal de Pernambuco
Centro de Informática
Recife, PE, Brazil
E-mail: mcps@cin.ufpe.br

Maricel G. Kann
University of Maryland/Baltimore County
Department of Biological Sciences
Baltimore, MD, USA
E-mail: mkann@umbc.edu

ISSN 0302-9743   e-ISSN 1611-3349
ISBN 978-3-642-31926-6   e-ISBN 978-3-642-31927-3
DOI 10.1007/978-3-642-31927-3
Springer Heidelberg Dordrecht London New York

Library of Congress Control Number: 2012942078

CR Subject Classification (1998): J.3, I.2, F.1, H.2.8, I.5, H.3

LNCS Sublibrary: SL 8 – Bioinformatics

Typesetting: Camera-ready by author, data conversion by Scientific Publishing Services, Chennai, India

Printed on acid-free paper

Springer is part of Springer Science+Business Media (www.springer.com)

Preface

This volume contains the papers selected for presentation at BSB 2012: the 7th Brazilian Symposium on Bioinformatics held August 15–17, 2012, in Campo Grande (Mato Grosso do Sul), Brazil. The BSB is an international conference which covers all aspects of bioinformatics and computational biology. The event is organized by the special interest group in Computational Biology of the Brazilian Computer Society (SBC). The BSB series started in 2005. In the period 2002–2004 its name was Brazilian Workshop on Bioinformatics (WOB). Thus, this year we celebrated the 10th anniversary.

As in previous editions, BSB 2012 had an international Program Committee of 55 members. After a rigorous review process, 16 papers were accepted to be orally presented. All papers were reviewed by at least three independent reviewers. This year, a selection of the accepted papers was also invited for submission in expanded format to a special issue of the *IEEE/ACM Transactions on Computational Biology and Bioinformatics (TCBB)*. We thank the IEEE/ACM and Marie-France Sagot, as editor-in-chief of the TCBB, for offering us this opportunity.

In addition to the technical presentations, BSB 2012 featured keynote talks. We are grateful to all of the invited speakers: Marcos V.G.B. da Silva (EMBRAPA, Brazil), José Fernando Garcia (UNESP, Brazil), Fernando D. González-Nilo (Universidad Andrés Bello, Chile), Maricel G. Kann (University of Maryland/Baltimore, USA), and Jens Stoye (Bielefeld University, Germany). These proceedings also include a joint paper from two of the guest speakers.

BSB 2012 was made possible by the dedication and work of many people and institutions, especially the *Faculdade de Computação* (Facom) of the *Universidade Federal do Mato Grosso do Sul* (UFMS). We would like to express our sincere thanks to all Program Committee members, as well as to the external reviewers, for their cooperation in the reviewing process. We appreciate their hard work that guaranteed the high quality of the technical program. We are also grateful to the local organizers, coordinated by Nalvo F. Almeida Jr. (Facom/UFMS, Brazil), and volunteers for their valuable help; the sponsors, in particular CNPq, CAPES and FUNDECT, for making the event financially viable; Guilherme Telles for assisting with the preparation of the proceedings; and Springer for agreeing to print this volume.

Finally, we would like to thank the authors for their time and effort in submitting their work to the BSB 2012, as well as the attendees. Without them this event would not be possible.

August 2012

Marcilio C.P. de Souto
Maricel G. Kann

Organization

BSB 2012 was supported by the Brazilian Computer Society (SBC) and organized by Faculdade de Computação da Universidade Federal de Mato Grosso do Sul (UFMS); Embrapa Gado de Corte; Fundação de Apoio ao Desenvolvimento do Ensino, Ciência e Tecnologia do Estado de Mato Grosso do Sul (Fundect); Museu das Culturas Dom Bosco.

Conference Chair

Nalvo Franco de Almeida Jr.   Universidade Federal de Mato Grosso do Sul, Brazil

Program Chairs

Marcilio C.P. de Souto   Universidade Federal de Pernambuco, Brazil
Maricel G. Kann   University of Maryland Baltimore County, USA

Steering Comittee

Marcelo de Macedo Brígido   Universidade de Brasília, Brazil
Ivan Gesteira Costa   Universidade Federal de Pernambuco, Brazil
Osmar Norberto de Sousa   Pontifícia Universidade Católica
do Rio Grande do Sul, Brazil
Carlos Eduardo Ferreira   Universidade de São Paulo, Brazil
João Carlos Setubal   Universidade de São Paulo, Brazil
Guilherme P. Telles   Universidade Estadual de Campinas, Brazil
Maria Emilia Walter   Universidade de Brasília, Brazil

Program Committee

Said S. Adi   Universidade Federal de Mato Grosso do Sul, Brazil
Nalvo F. Almeida Jr.   Universidade Federal de Mato Grosso do Sul, Brazil
Fernando Álvarez-Valín   Universidad de la Republica, Uruguay
Ana L.C. Bazzan   Universidade Federal do Rio Grande do Sul, Brazil
Ana M. Benko-Iseppon   Universidade Federal de Pernambuco, Brazil
Marília D.V. Braga   INMETRO, Brazil
Marcelo de Macedo Brígido   Universidade de Brasília, Brazil

Additional Reviewers

Paulo Alvarez	Universidade de Brasília, Brazil
Ronnie Alves	Universidade Federal do Rio Grande do Sul, Brazil
George Cavalcanti	Universidade Federal de Pernambuco, Brazil
Guillaume Cleuziou	Université d'Orléans, France
André Coelho	Universidade de Fortaleza, Brazil
Bruno Motta de Carvalho	Universidade Federal do Rio Grande do Norte, Brazil
Marcelo Henriques de Carvalho	Universidade Federal de Mato Grosso do Sul, Brazil
Daniel de Oliveira	Universidade Federal do Rio de Janeiro, Brazil
Katti Faceli	Universidade Federal de São Carlos/Sorocaba, Brazil
Cristina Fernandes	Universidade de São Paulo, Brazil
Celina Figueiredo	Universidade Federal do Rio de Janeiro, Brazil
Alvaro Franco	Universidade de São Paulo, Brazil
Alexandre Freire	Universidade de São Paulo, Brazil
Chih-Hao Hsu	NIH, USA
Andre Kashiwabara	Universidade Tecnológica Federal do Paraná, Brazil
Marta Kwiatkowska	University of Oxford, UK
Ivani Lopes	Embrapa, Brazil
Fábio Martinez	Universidade Federal de Mato Grosso do Sul, Brazil
Karina S. Machado	Universidade Federal do Rio Grande, Brazil
Leandro Marzulo	Universidade do Estado do Rio de Janeiro, Brazil
Mariana Mendoza	Universidade Federal do Rio Grande do Sul, Brazil
Luciana Montera	Universidade Federal de Mato Grosso do Sul, Brazil
Deborah Muganda	University of Wisconsin, USA
Alexandre Noma	Universidade Federal do ABC, Brazil
Kary Ocaña	Universidade Federal do Rio de Janeiro, Brazil
Thiago Lipinski Paes	Pontifícia Universidade Católica do Rio Grande do Sul, Brazil
Ronaldo Prati	Universidade Federal do ABC, Brazil
Paula Andreia Silva	Universidade Católica de Brasília, Brazil
Renata Souza	Universidade Federal de Pernambuco, Brazil
Jens Stoye	Universität Bielefeld, Germany
Renato Tinós	Universidade de São Paulo/Ribeirão Preto, Brazil
Adriano Werhli	Universidade Federal do Rio Grande, Brazil
Cleber Zanchettin	Universidade Federal de Pernambuco, Brazil

Financial Sponsors

Conselho Nacional de Desenvolvimento Científico e Tecnológico (CNPq)
Coordenação de Aperfeiçoamento de Pessoal de Nível Superior (CAPES)
Fundação de Apoio ao Desenvolvimento do Ensino, Ciência e Tecnologia do Estado de Mato Grosso do Sul (Fundect)
Fundação de Apoio à Pesquisa ao Ensino e à Cultura (FAPEC)
Unidade Especial de Criação e Inovação da Prefeitura Municipal de Campo Grande
Governo do Estado de Mato Grosso do Sul
Life Technologies
Copagaz
H2L Soluções para Documentos
Netware Enterprise
Alfa Computadores
Tecsinapse
Paulistão
Infor-7

Table of Contents

Transposition Diameter and Lonely Permutations

Luís Felipe I. Cunha[1], Luis Antonio B. Kowada[2],
Rodrigo de A. Hausen[3], and Celina M.H. de Figueiredo[1]

[1] Universidade Federal do Rio de Janeiro
{lfignacio,celina}@cos.ufrj.br
[2] Universidade Federal Fluminense
luis@vm.uff.br
[3] Universidade de São Paulo
hausen@compscinet.org

Abstract. Determining the transposition distance of permutations was proven recently to be NP-hard. However, the problem of the transposition diameter is still open. The known lower bounds for the diameter were given by Meidanis, Walter and Dias when the lengths of the permutations are even and by Elias and Hartman when the lengths are odd. A better lower bound for the transposition diameter was proposed using the new definition of *super-bad permutations*, that would be a particular family of the *lonely permutations*. We show that there are no super-bad permutations, by computing the exact transposition distance of the union of two copies of particular lonely permutations that we call *knot permutations*. Meidanis, Walter, Dias, Elias and Hartman, therefore, still hold the current best lower bound. Moreover, we consider the union of distinct lonely permutations and manage to define an alternative family of permutations that meets the current lower bound.

Keywords: comparative genomics, genome rearrangement, transposition diameter, lonely permutations, knot permutations.

1 Introduction

By means of comparing the orders of common genes between two organisms, one may estimate the series of mutations that occurred in the underlying evolutionary process. In the simplified genome rearrangement model adopted in this paper, each mutation is a transposition, and the sole chromosome of each organism is modeled by a permutation, which means that there are no duplicated or deleted genes. A transposition is a rearrangement of the gene order within a chromosome, in which two contiguous blocks are swapped. A biological explanation for this rearrangement is the duplication of a block of genes, followed by the deletion of the original block [2,14].

The transposition distance is the minimum number of transpositions required to transform one chromosome into the other. Bulteau *et al.* proved that the

M.C.P. de Souto and M.G. Kann (Eds.): BSB 2012, LNBI 7409, pp. 1–12, 2012.

problem of determining the transposition distance between two permutations is NP-hard [3]. Nevertheless, the transposition diameter, i.e. the largest possible value of the transposition distance of the symmetric group S_n, is still an open problem. Meidanis *et al.* [13] showed the transposition distance $\lfloor \frac{n}{2} \rfloor + 1$ for the reverse permutation and conjectured that it was equal to the transposition diameter. Eriksson *et al.* [7] showed two examples of permutations with odd number of elements such that the transposition distance is more distant than the reverse, and Elias and Hartman [6], later, showed the lower bound of $\lfloor \frac{n+1}{2} \rfloor + 1$ by constructing permutations with odd number of elements based upon the examples of [7].

Lu and Yang [12] proposed an improvement to the lower bound by postulating that, if there existed a special hard-to-sort permutation that satisfied a couple of properties — a so-called *super-bad permutation* — then it would be possible to construct an infinite family of permutations whose distance would be greater than $\lfloor \frac{n+1}{2} \rfloor + 1$. The construction is based on the *union operation*, as described in [4,6]. A consequence of the definition of super-bad is that the only permutations to be super-bad would be one particular family of the known lonely permutations [9,10], that we call *knot permutation*. However, we present a proof that the super-bad permutations do not exist, and consequently, the best lower bounds on the transposition diameter are still given in [6,13].

This article is organized as follows: Section 2 provides the basic background on lonely permutations, transposition distance, and transposition diameter; the structure introduced by Bafna and Pevzner named breakpoint graph, and the particularities on lonely permutations; and we also present the known bounds on the transposition diameter. Section 3 is devoted to demonstrate the nonexistence of the super-bad permutations, and Section 4 concludes the paper by considering other unions of lonely permutations.

2   Background

For our propose, we assume a chromosome as a permutation and each gene, without repetition on the chromosome, as a positive distinct integer.

Definition 1. *[1] A permutation $\pi = [\pi_1 \pi_2 \ldots \pi_n]$ is the bijective function of $[n] = \{1, 2, \ldots, n\}$ onto itself such that $\pi(i) = \pi_i$ for $1 \leq i \leq n$.*

Remark that any permutation of length n is an element of the symmetric group S_n, i.e. the product of two permutations π and σ of length n is also a permutation with n elements, each permutation π has one associated inverse permutation π^{-1} such that the product $\pi \cdot \pi^{-1} = \iota_{[n]} = [1\,2\,\cdots\,n-1\,n]$ and for any permutation π of length n we have that $\pi \cdot \iota_{[n]} = \pi$. The symbol $\iota_{[n]}$ denotes the *identity permutation* and the symbol $\rho_{[n]}$ denotes the *reverse permutation* $[n\,n{-}1\,\cdots\,2\,1]$.

Definition 2. *[1] The transposition denoted by $t(i, j, k)$, where $1 \leq i < j < k \leq n+1$, is the permutation $[1\,2\cdots i{-}1\,j\,j{+}1\cdots k{-}1\,i\,i{+}1\cdots j{-}1\,k\,\cdots n]$.*

The product of π and $t(i,j,k)$ is an application of the transposition $t(i,j,k)$ to π, as an action to the right, i. e. $\pi \cdot t(i,j,k)$ is the permutation

$$\left[\pi_1 \ \pi_2 \ \cdots \ \pi_{i-1} \boxed{\pi_j \ \cdots \ \pi_{k-1}} \boxed{\pi_i \ \cdots \ \pi_{j-1}} \pi_k \ \cdots \ \pi_n\right].$$

Definition 3. *[1] The transposition distance $d_t(\pi,\sigma)$ of a permutation π with respect to the permutation σ, where π and σ have the same number of elements, is the length q of a shortest sequence of transpositions $t_1, t_2, \cdots, t_q$, such that $\pi t_1 t_2 \cdots t_q = \sigma$. If $\pi = \sigma$, then $d_t(\pi,\sigma) = 0$. The transposition diameter TD(n) is the maximum transposition distance among all permutations belonging to the symmetric group S_n.*

Since $d_t(\pi,\sigma) = d_t(\pi\sigma^{-1}, \iota)$, we may consider $\sigma = \iota$ and denote $d_t(\pi) = d_t(\pi,\iota)$.

Permutations can be joined into the so-called toric equivalence classes; it comes as a consequence of the definition of such classes that permutations in the same class have the same transposition distance with respect to the identity [7]. However, some toric classes have just one permutation, so they are called *unitary* toric classes and correspond to the *lonely permutations* [11].

Theorem 1. *[9,10] A permutation π is a lonely permutation, if and only if, $\pi = [\overline{\ell} \ \overline{2\ell} \ \overline{3\ell} \ \cdots \ \overline{n\ell}]$ for some $\ell \geq 1$, where $gcd(n+1,\ell) = 1$ and $\overline{x}$ is the remainder of the division of x by $n+1$.*

We denote lonely permutations by $u_{n,\ell}$, where $n+1$ and ℓ are coprime. Remark that $\rho_{[n]} = u_{n,n}$ and $\iota_{[n]} = u_{n,1}$. We have considered lonely permutations as candidates to investigate the transposition diameter TD(n) [10,11].

Example 1. A possible lonely permutation with $n = 10$ is when $\ell = 8$, and it is: $u_{10,8} = [8\,5\,2\,10\,7\,4\,1\,9\,6\,3]$.

A structure given by Bafna and Pevzner [1], which allowed the development of non-trivial bounds on the transposition distance, is presented next.

The Breakpoint Graph

Definition 4. *[1] Given a permutation π, the breakpoint graph of π, denoted by $G(\pi)$, is a multigraph on the set of vertices $V = \{0, -1, +1, -2, +2, \cdots, -n, +n, -(n+1)\}$, whose set of edges is the union of two disjoint sets: $D = \{(+i, -(i+1)) \ |i = 0, \cdots, n\}$, of the so-called desire edges, and $R = \{(+\pi_i, -\pi_{i+1}) \ |i = 1, \cdots, n-1\} \cup \{(0, -\pi_1), (+\pi_n, -(n+1))\}$, of the reality edges[1].*

As a direct consequence of the definition, every vertex in $G(\pi)$ has degree two, so $G(\pi)$ can be partitioned into disjoint cycles. We shall use the terms *a cycle in π* and *a cycle in $G(\pi)$* interchangeably to denote the latter. We say that a cycle in π has length k, or that it is a k-cycle, if it has exactly k reality edges (or, equivalently, k desire edges).

---

[1] The reality edges in $G(\pi)$ represent the adjacencies between the elements in π and the desire edges represent the adjacencies between the elements in $\iota_{[n]}$.

The number of cycles of odd length in $G(\pi)$, denoted by $c_{odd}(\pi)$, is an important measure. Bafna and Pevzner [1] proved that, after applying a transposition to a permutation, the number of odd cycles in its breakpoint graph changes in one of the following ways: i) it *increases* by two units; ii) it does not change; or iii) it *decreases* by two units. Every transposition can be classified according to its effect on the number of odd cycles into a: i) 2-move; ii) 0-move; or iii) −2-move, respectively. This classification results in the lower bound in Theorem 2.

Theorem 2. *[1] The transposition distance is* $d_t(\pi) \geq \left\lceil \frac{(n+1)-c_{odd}(\pi)}{2} \right\rceil$.

Meidanis, Walter and Dias [13] observed that, for $n \geq 3$, whenever $n \equiv 3$ (mod 4), the number of odd cycles in $\rho_{[n]}$ is 0, which gives a lower bound of $d_t(\rho) \geq \frac{n+1}{2} = \lfloor \frac{n}{2} \rfloor + 1$. For values of $n \geq 3$ other than those, they have found that the first two consecutive transpositions applied to $\rho_{[n]}$ cannot be both a 2-move; therefore, the lower bound $d_t(\rho) \geq \lfloor \frac{n}{2} \rfloor + 1$ is also valid. These facts, coupled with the fact that it is always possible to find a sequence of $\lfloor \frac{n}{2} \rfloor + 1$ transpositions that transforms $\rho_{[n]}$ into $\iota_{[n]}$, results in Theorem 3.

Theorem 3. *[13] The transposition distance of a reverse permutation of length* $n \geq 3$ *is* $d_t(\rho_{[n]}) = \lfloor \frac{n}{2} \rfloor + 1$.

Reductions and the Order of Elements in a Cycle

Definition 5. *[4] The* reduced permutation *of* π*, denoted* $gl(\pi)$*, is the permutation whose breakpoint graph* $G(gl(\pi))$ *is equal to* $G(\pi)$ *without the cycles of length 1, and having its vertices relabeled accordingly.*

Definition 6. *[4] Two permutations* π *and* σ *are* equivalent by reduction*, if* $gl(\pi) = gl(\sigma)$.

Definition 7. *[11] A permutation* σ *is an* r-reduction *of* π *if there is a sequence of* r *transpositions, all of them 2-moves, that transforms* π *into a permutation that is equivalent by reduction to* σ.

Corollary 1. *[11] If* σ *is an* r-reduction *of* π*, then* $d_t(\pi) \leq d_t(\sigma) + r$.

It is possible to represent a permutation π by the sequence of the non-negative elements in every cycle of $G(\pi)$ or by the corresponding positions of the elements on the permutation.

Definition 8. *Given a permutation* π*, the sequence of non-negative elements, starting from the leftmost element, in each cycle* i *that contains the element* i *of* $G(\pi)$ *is denoted by* $c_i(\pi)$*, and a collection of those sequences is denoted by* $C(\pi)$.

We use parentheses to delimit the cycle sequences of $C(\pi)$*. The number of cycles in the breakpoint graph* $G(\pi)$ *is* $|C(\pi)|$.

Definition 9. *Given a permutation π and each sequence of non-negative elements $c_i(\pi)$, the corresponding sequence of positions of the respective elements is denoted by $p_i(\pi)$. The collection of the sequences of positions of non-negative elements is denoted by $P(\pi)$.*

We use angle brackets — "$\langle$" and "$\rangle$" — to delimit the corresponding sequences of positions.

If a permutation π has just one cycle we use $c(\pi)$ and $p(\pi)$ to denote $c_0(\pi)$ and $p_0(\pi)$, respectively. Fig. 1 illustrates how the cycle sequences in $C(\pi)$ and the sequences of positions in $P(\pi)$ relate to $G(\pi)$.

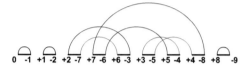

Fig. 1. $G(\pi)$ for $\pi = [1\,2\,7\,6\,3\,5\,4\,8]$. $C(\pi) = \{(0),(1),(2\,6),(7\,5\,3\,4),(8)\}$ and $P(\pi) = \{\langle 0\rangle, \langle 1\rangle, \langle 2\,4\rangle, \langle 3\,6\,5\,7\rangle, \langle 8\rangle\}$. Remark that $c_2 = c_6$ and $c_7 = c_5 = c_3 = c_4$.

The Knot Permutation u_{n,ℓ^*}

Lemma 1. *[11] Let $u_{n,\ell}$ be a lonely permutation, with $\ell > 1$. Then $G(u_{n,\ell})$ satisfies:*

1. *each cycle, $c_i(u_{n,\ell})$, has length $k = (n+1)/\gcd(n+1,\ell-1)$;*
2. *the number of cycles, $|C(u_{n,\ell})|$, is $\gcd(n+1,\ell-1)$;*
3. *each cycle $c_i(u_{n,\ell}) = (+i, +\overline{i+\ell-1}, +\overline{i+2(\ell-1)}, \ldots, +\overline{i+(k-1)(\ell-1)})$, for $i = 0, \ldots, |C(u_{n,\ell})| - 1$.*

Lemma 2. *[11] Consider the cycle that contains the vertex 0, $c_0(u_{n,\ell}) = (0, +\overline{\ell-1}, +\overline{2(\ell-1)}, \ldots)$. Then $p_0(u_{n,\ell}) = \langle 0, m, \overline{2m}, \ldots\rangle$, where ℓ^{-1} is an integer such that $\ell\ell^{-1} \equiv 1\,(\mathrm{mod}\,n+1)$ and $m \equiv 1 - \ell^{-1}\,(\mathrm{mod}\,n+1)$.*

Christie [4] defined permutations denoted by ω, with just one cycle called *knot*, and with $2r$ elements.

Definition 10. *[4] The permutation ω is the permutation such that $p(\omega) = \langle 0, \frac{n}{2}, n, \frac{n}{2} - 1, n - 1, \frac{n}{2} - 2, n - 2, \cdots, 1, n - (\frac{n}{2} - 1)\rangle$.*

Theorem 4. *[4] The permutation ω with $2r$ elements only exists when $r \in \{3q, 3q - 1\}$ for some integer $q \geq 1$. Moreover, ω is the unique permutation that satisfies: i) ω has just one cycle; ii) there is no 2-move that can be applied on ω.*

Definition 11. *The* knot permutation *denoted u_{n,ℓ^*} is such that:*

$$\ell^* = \begin{cases} 2q + 1 & \text{for } n = 6q, \\ 4q & \text{for } n = 6q - 2. \end{cases}$$

Note that $gcd(6q + 1, 2q + 1) = 1$ and $gcd(6q - 1, 4q) = 1$. Definition 11 was first proposed in [10] assuming that $6q + 1$ and $6q - 1$ are prime.

Corollary 2. *The knot permutation $u_{n,\ell*}$ is the ω permutation.*

Notice that ω and the knot permutation are defined for the same values of n. Comparing Lemma 2 with Definition 10 we have that $p(u_{n,\ell*}) = p(\omega)$.

Theorem 5. *[9,10] The transposition distance of the knot permutation: $u_{n,\ell*}$ is $d_t(u_{n,\ell*}) \geq \frac{n}{2} + 1$.*

Independently, Christie [4] showed the sequence of $n/2 + 1$ transpositions needed to sort $u_{n,\ell*}$, so $d_t(u_{n,\ell*}) = \frac{n}{2} + 1$.

3   Nonexistence of Super-Bad Permutations

Definition 12. *[4,6,12] Given $\pi \in S_n$ and $\sigma \in S_m$, the union of π and σ is an element $\pi \uplus \sigma \in S_{n+m+1}$ such that:*

$$\pi \uplus \sigma = [\pi_1 \cdots \pi_{n-1} \pi_n (n+1) (n+1+\sigma_1) \cdots (n+1+\sigma_{m-1}) (n+1+\sigma_m)].$$

The union operation $\uplus$ was first used with the symbol $++$ by Christie [4] to construct permutations that are distant from the lower bound on the transposition distance given by Theorem 2. Elias and Hartman [6] later used the union operation to construct two families of permutations corresponding to the current lower bound on the transposition diameter for odd n.

The first non-trivial lower bound on the transposition diameter was given by Meidanis, Walter and Dias [13], as a consequence of Theorem 3. They conjectured that $\lfloor n/2 \rfloor + 1$ was the exact transposition diameter. However, Eriksson et al. [7] found two permutations, $[4\,3\,2\,1\,5\,13\,12\,11\,10\,9\,8\,7\,6] = \rho_{[4]} \uplus \rho_{[8]}$ and $[4\,3\,2\,1\,5\,15\,14\,13\,12\,11\,10\,9\,8\,7\,6] = \rho_{[4]} \uplus \rho_{[10]}$, for which the transposition distance was 8 and 9 respectively, invalidating the conjecture for TD(13) and TD(15).

A better lower bound for the diameter for odd values of $n \geq 13$ was given by Elias and Hartman [6]. Building upon the permutations found by Eriksson et al. [7], they constructed families of permutations with odd number of elements such that their transposition distance are $\lfloor (n+1)/2 \rfloor + 1$. For $n = 13 + 4k$, they considered the permutation $((((\rho_{[4]} \uplus \rho_{[8]}) \underbrace{\uplus \rho_{[3]}) \uplus \rho_{[3]}) \dots \uplus \rho_{[3]}}_{k \text{ repetitions of } \rho_{[3]}})$, and for $n =$

$15 + 4k$, the permutation $((((\rho_{[4]} \uplus \rho_{[10]}) \underbrace{\uplus \rho_{[3]}) \uplus \rho_{[3]}) \dots \uplus \rho_{[3]}}_{k \text{ repetitions of } \rho_{[3]}})$. Remark that, for

permutations with an even number of elements, so far there are no permutations whose transposition distance exceeds that of the reverse permutation.

Lu and Yang [12] proposed an improvement for the lower bound using unions of *super-bad permutations*.

Definition 13. *[12] The permutation γ is a super-bad permutation if: i) γ has just one cycle; ii) there is no 2-move that can be applied on γ; iii) every sequence of transpositions that sort $\gamma \uplus \gamma$ contains at least two 0-moves.*

As shown in Corollary 2, the knot permutations u_{n,ℓ^*} is the only family satisfying conditions i and ii. Additionally, for u_{n,ℓ^*} to be a super-bad permutation, $u_{n,\ell^*} \uplus u_{n,\ell^*}$ would have transposition distance at least $n+2$. However, Theorem 6 below shows that $d_t(u_{n,\ell^*} \uplus u_{n,\ell^*}) = n + 1$, and so we guarantee the nonexistence of super-bad permutations.

We demonstrate that, for every u_{n,ℓ^*}, there exists a sequence that sorts $u_{n,\ell^*} \uplus u_{n,\ell^*}$ with just one 0-move. The first three transpositions allow us to reduce the problem, from $2n+1$ to $2n-3$, as shown in Lemma 3, which follows Definitions 14 and 15.

Definition 14. *[8] The concatenation of a sequence a with a sequence b, denoted by $a \odot b$, is the operation that joins both sequences. The generalized concatenation is denoted by $\bigodot_{x=y}^{z} f(x)$ and is the concatenation of the sequences $f(x)$, with x ranging from y to z.*

Definition 15. *Let $\delta_{[12q-3]}$ and $\epsilon_{[12q-7]}$, where $q \geq 3$, be the permutations defined as follows:*

$$\delta_{[12q-3]} = \bigodot_{i=2}^{q}[(2q + 2i - 3)(4q + 2i - 3)(2i - 2)] \odot [(4q - 1)(10q - 1)(6q)]\odot$$
$$\odot \bigodot_{i=2}^{q}[(8q + 2i - 4)(10q + 2i - 3)(6q + 2i - 2)] \odot [2q]\odot$$
$$\odot \bigodot_{i=q+2}^{2q}[(2q + 2i - 4)(2i - 2q - 3)(2i - 2)] \odot [(6q - 2)(2q - 1)]\odot$$
$$\odot \bigodot_{i=q+1}^{2q}[(8q + 2i - 4)(4q + 2i - 3)(6q + 2i - 3)], \quad and$$

$$\epsilon_{[12q-7]} = \bigodot_{i=2}^{q}[(4q + 2i - 5)(2q + 2i - 4)(2i - 2)] \odot [(8q - 3)(6q - 2)]\odot$$
$$\odot \bigodot_{i=2}^{q-1}[(10q + 2i - 7)(8q + 2i - 5)(2i + 6q - 4)]\odot$$
$$\odot[(12q - 7)(10q - 5)(4q - 2)(2q - 1)]\odot$$
$$\odot \bigodot_{i=q+2}^{2q-1}[(2i - 2q - 3)(2q + 2i - 4)(2i - 3)]\odot$$
$$\odot[(2q - 3)(6q - 4)(4q - 3)(8q - 4)]\odot$$
$$\odot \bigodot_{i=q+1}^{2q-1}[(2i + 4q - 5)(8q + 2i - 6)(2i + 6q - 4)] \odot [8q - 5].$$

Lemma 3. *Given n such that u_{n,ℓ^*} is a knot permutation, consider the transpositions $t_1 = t(1,n/2+2,n+2)$, $t_2 = t(2,n/2+3,n+3)$ and $t_3 = t(n/2+1,n+2,3n/2+2)$. For $q \geq 3$, if $n = 6q$ then $gl((u_{n,\ell^*} \uplus u_{n,\ell^*})t_1 t_2 t_3) = \delta_{[12q-3]}$, whereas if $n = 6q-2$ then $gl((u_{n,\ell^*} \uplus u_{n,\ell^*})t_1 t_2 t_3) = \epsilon_{[12q-7]}$.*

Proof. By constructing $u_{n,\ell^*} \uplus u_{n,\ell^*}$, we have that it is impossible to apply a 2-move on the first transposition, each cycle has the structure of the knot permutation, and since each cycle has odd length, if the transposition joins the two cycles it does not increase the number of odd cycles.

For the case $n = 6q$, let $\alpha = u_{6q,2q+1} \uplus u_{6q,2q+1}$. Therefore, α can be described as the following concatenation:

$$\alpha = \bigodot_{i=1}^{q}[(2q + 2i - 1)(4q + 2i)(2i)]\odot$$
$$\odot \bigodot_{i=q+1}^{2q}[(2q + 2i - 1)(2i - 2q - 1)(2i)]\odot$$
$$\odot[6q + 1] \odot \bigodot_{i=1}^{q}[(8q + 2i)(10q + 2i + 1)(6q + 2i + 1)]\odot$$
$$\odot \bigodot_{i=q+1}^{2q}[(8q + 2i)(4q + 2i)(6q + 2i + 1)].$$

Applying the transpositions t_1, t_2 and t_3 to α, in this order, yields:

$$\alpha''' = [(1)\,(2)] \odot \bigodot_{i=2}^{q}[(2q+2i-1)\,(4q+2i)\,(2i)] \odot [(4q+1)\,(4q+2)]\odot$$
$$\odot[(10q+3)\,(6q+3)] \odot \bigodot_{i=2}^{q}[(8q+2i)\,(10q+2i+1)\,(6q+2i+1)]\odot$$
$$\odot[(8q+2)\,(2q+2)] \odot \bigodot_{i=q+2}^{2q}[(2q+2i-1)\,(2i-2q-1)\,(2i)]\odot$$
$$\odot[(6q+1)\,(2q+1)] \odot \bigodot_{i=q+1}^{2q}[(8q+2i)\,(4q+2i)\,(6q+2i+1)].$$

The reduction of α''' is obtained by eliminating the elements 1, 2, $4q+2$ and $8q+2$, with the subsequent renumbering of the remaining elements. This results in $\delta_{[12q-3]}$.

For the case $n = 6q-2$, let $\beta = u_{6q-2,4q} \uplus u_{6q-2,4q}$. Therefore, β can be described as the following concatenation:

$$\beta = \bigodot_{i=1}^{q}[(4q+2i-2)\,(2q+2i-1)\,(2i)]\odot$$
$$\odot\bigodot_{i=q+1}^{2q-1}[(2i-2q-1)\,(2q+2i-1)\,(2i)] \odot [(2q-1)\,(6q-1)]\odot$$
$$\odot\bigodot_{i=1}^{q}[(10q+2i-3)\,(8q+2i-2)\,(2i+6q-1)]\odot$$
$$\odot\bigodot_{i=q+1}^{2q-1}[(2i+4q-2)\,(8q+2i-2)\,(2i+6q-1)] \odot [8q-2].$$

Applying the transpositions t_1, t_2 and t_3 to β, in this order, yields:

$$\beta''' = [(1)\,(2)] \odot \bigodot_{i=2}^{q}[(4q+2i-2)\,(2q+2i-1)\,(2i)] \odot [(2q+1)\,(8q)]\odot$$
$$\odot[(6q+1)] \odot \bigodot_{i=2}^{q-1}[(10q+2i-3)\,(8q+2i-2)\,(2i+6q-1)]\odot$$
$$\odot[(12q-3)\,(10q-2)\,(10q-1)\,(4q+1)\,(2q+2)]\odot$$
$$\odot\bigodot_{i=q+2}^{2q-1}[(2i-2q-1)\,(2q+2i-1)\,(2i)]\odot$$
$$\odot[(2q-1)\,(6q-1)\,(4q)\,(8q-1)]\odot$$
$$\odot\bigodot_{i=q+1}^{2q-1}[(2i+4q-2)\,(8q+2i-2)\,(2i+6q-1)] \odot [8q-2].$$

The reduction of β''' is obtained by eliminating the elements 1, 2, $2q+1$ and $10q-1$, with the subsequent renumbering of the remaining elements. This results in $\epsilon_{[12q-7]}$.   □

Lemma 4 verifies that the permutations $u_{4,4} \uplus u_{4,4}$, $u_{6,3} \uplus u_{6,3}$, $u_{10,8} \uplus u_{10,8}$ and $u_{12,5} \uplus u_{12,5}$ can be sorted with only one 0-move, for after applying the same transpositions t_1, t_2 and t_3, in this order, specified in Lemma 3 one obtains permutations that can be sorted using just 2-moves.

Lemma 4. *For $q = 1$ and $q = 2$ the permutations $gl((u_{n,\ell^*} \uplus u_{n,\ell^*})t_1t_2t_3)$, where $n = 6q$ or $n = 6q-2$, and $t_1 = t(1,n/2+2,n+2)$, $t_2 = t(2,n/2+3,n+3)$ and $t_3 = t(n/2+1,n+2,3n/2+2)$, can be sorted applying $n-2$ transpositions.*

Proof. Let us consider the cases when $q = 1$ and $q = 2$ after applying the transpositions t_1, t_2 and t_3 and reducing the produced permutation. If $q = 1$ and $n = 4$, we apply $t(1,3,5)$ and $t(2,4,6)$; else we apply $t(1,n/2-1,n-1)$, $t(2,n/2,3n/2)$ and if $q = 1$ and $n = 6$ we apply $t(1,5,7)$, $t(2,4,10)$; else we apply $t(n,3n/2-1,2n-2)$, $t(n/2,n-1,3n/2)$, $t(2,n/2+3,3n/2+2)$, $t(1,n/2+2,n+3)$ and if $q = 2$ and $n = 10$ we apply $t(5,9,15)$ and $t(2,7,13)$; if $q = 2$ and $n = 12$ we apply $t(n-1,3n/2,2n-3)$, $t(n/2-3,3n/2-2,2n-5)$, $t(9,13,21)$, $t(2,15,19)$.

Therefore, Lemma 4 establishes the basic cases: no knot u_{n,ℓ^*} with $n \leq 12$ can be a super-bad permutation. Lemma 5 and Theorem 6 extend this conclusion to every knot with more than 12 elements.

Lemma 5. *For any $q \geq 3$ the permutation $\delta_{[12(q-2)-3]}$ is a 12-reduction of the permutation $\delta_{[12q-3]}$, $d_t(\delta_{[12q-3]}) = d_t(\delta_{[12(q-2)-3]}) + 12$. And the permutation $\epsilon_{[12(q-2)-7]}$ is a 12-reduction of the permutation $\epsilon_{[12q-7]}$, $d_t(\epsilon_{[12q-7]}) = d_t(\epsilon_{[12(q-2)-7]}) + 12$.*

Proof. Let q be an integer greater than 2. If the permutation to be sorted is $\delta_{[12q-3]}$, consider $n = 12q - 3$, whereas if $\epsilon_{[12q-7]}$ is to be sorted, consider $n = 12q - 7$. The transpositions that will be applied in both cases are

$$
\begin{aligned}
&t_1 = t(1, \tfrac{n}{2}-1, n-1) & &t_5 = t(2, \tfrac{n}{2}+3, \tfrac{3n}{2}+2) & &t_9 = t(n-3, n+1, \tfrac{3n}{2}-1) \\
&t_2 = t(2, \tfrac{n}{2}, \tfrac{3n}{2}) & &t_6 = t(1, \tfrac{n}{2}+1, n+3) & &t_{10} = t(2, \tfrac{n}{2}, \tfrac{3n}{2}-3) \\
&t_3 = t(n, \tfrac{3n}{2}-1, 2n-2) & &t_7 = t(n-1, \tfrac{3n}{2}, 2n-3) & &t_{11} = t(\tfrac{n}{2}-5, \tfrac{3n}{2}-7, 2n-5) \\
&t_4 = t(\tfrac{n}{2}, n-1, \tfrac{3n}{2}) & &t_8 = t(\tfrac{n}{2}-3, \tfrac{3n}{2}-2, 2n-5) & &t_{12} = t(n-6, \tfrac{3n}{2}-5, 2n-3)
\end{aligned}
$$

For the case $\delta_{[12q-3]}$, after applying $t_1 \ldots t_{12}$ in this order, we obtain σ'^{12}:

$$
\begin{aligned}
\sigma'^{12} = &[(1)(2)] \odot \bigodot_{i=3}^{q-1}[(2q+2i-3)(4q+2i-3)(2i-2)] \odot [4q-3] \odot \\
&\odot[(4q-2)(4q-1)(4q)(4q+1)(10q-1)(10q)(10q+1)(6q+2)] \odot \\
&\odot \bigodot_{i=3}^{q-1}[(8q+2i-4)(10q+2i-3)(6q+2i-2)] \\
&\odot[(8q-3)(8q-2)(8q-1)(8q)(2q)(2q+1)(2q+2)] \odot \\
&\odot \bigodot_{i=q+3}^{2q-1}[(2q+2i-4)(2i-2q-3)(2i-2)] \\
&\odot[(6q-4)(2q-3)(2q-2)(2q-1)] \odot \\
&\odot[(10q-4)(10q-3)(10q-2)(6q-3)(6q-2)(6q-1)(6q)] \odot \\
&\odot[(6q+1)(8q+1)] \odot \bigodot_{i=q+2}^{2q-1}[(8q+2i-4)(4q+2i-3) \\
&(6q+2i-3)] \odot [(12q-4)(12q-3)].
\end{aligned}
$$

The reduction of σ'^{12} is obtained by eliminating the elements 1, 2, $4q-2$, $4q-1$, $4q$, $4q+1$, $10q$, $10q+1$, $8q-3$, $8q-2$, $8q-1$, $8q$, $2q+1$, $2q+2$, $2q-2$, $2q-1$, $10q-3$, $10q-2$, $6q-2$, $6q-1$, $6q$, $6q+1$, $12q-4$ and $12q-3$, with the subsequent renumbering of the remaining elements. This results in $\delta_{[12(q-2)-3]}$.

For the case $\epsilon_{[12q-7]}$, after applying $t_1 \ldots t_{12}$ in this order, we obtain σ''^{12}:

$$
\begin{aligned}
\sigma''^{12} = &[(1)(2)] \odot \bigodot_{i=3}^{q-1}[(4q+2i-5)(2q+2i-4)(2i-2)] \odot [(2q-3)] \odot \\
&\odot[(2q-2)(2q-1)(2q)(8q-3)(8q-2)(8q-1)(6q)] \odot \\
&\odot \bigodot_{i=3}^{q-2}[(10q+2i-7)(8q+2i-5)(2i+6q-4)] \odot [12q-9] \odot \\
&\odot[(10q-7)(10q-6)(10q-5)(10q-4)(10q-3)(4q-2)] \odot \\
&\odot[(4q-1)(4q)(2q+1)] \odot \\
&\odot \bigodot_{i=q+3}^{2q-1}[(2i-2q-3)(2q+2i-4)(2i-3)] \odot [(4q-4)(4q-3)] \odot \\
&\odot[(8q-6)(8q-5)(8q-4)(6q-5)(6q-4)(6q-3)(6q-2)] \odot \\
&\odot \bigodot_{i=q+2}^{2q-2}[(2i+4q-5)(8q+2i-6)(2i+6q-4)] \odot \\
&\odot[(8q-7)(12q-8)(12q-7)].
\end{aligned}
$$

The reduction of σ''^{12} is obtained by eliminating the elements 1, 2, $2q-3$, $2q-2$, $2q-1$, $2q$, $8q-2$, $8q-1$, $10q-6$, $10q-5$, $10-4$, $10q-3$, $4q-1$, $4q$, $4q-4$, $4q-3$, $8q-5$,

$8q-4$, $6q-4$, $6q-3$, $6q-2$, $6q-1$, $12q-8$, $12-7$, with the subsequent renumbering of the remaining elements. This results in $\epsilon_{[12(q-2)-7]}$. $\square$

Theorem 6. *The transposition distance of $u_{n,\ell^*} \uplus u_{n,\ell^*}$ is equal to $n+1$. Therefore, no super-bad permutation exists.*

Proof. The permutation $u_{n,\ell^*} \uplus u_{n,\ell^*}$ has $2n+1$ elements and 2 odd cycles, which gives us a lower bound of $\frac{(2n+1+1)-2}{2} = n$ for its transposition distance, according to Theorem 2. However, the first transposition cannot be a 2-move, as we showed in Lemma 3, so the transposition distance is at least $n+1$.

Lemma 3 states that it is possible to apply one 0-move and two 2-moves in order to obtain a permutation that is equivalent by reduction to $\delta_{[12q-3]}$, or to $\epsilon_{[12q-7]}$; since both cases are similar, we will only write down the reasoning for $\delta_{[12q-3]}$.

If $q \leq 2$, we have already seen that, in those basic cases, $(u_{n,\ell^*} \uplus u_{n,\ell^*})t_1t_2t_3$ can be sorted using only 2-moves, Lemma 4. Otherwise, if $q \geq 3$, Lemma 5 is repeatedly applied i times, obtaining $\delta_{[12(q-2)-3]}, \delta_{[12(q-4)-3]}, \cdots, \delta_{[12(q-2i)-3]}$, using only 2-moves, until $q - 2i \leq 3$, and then $gl((\delta_{[12(q-2i)-3]})t_{12})$, with $t_{12} = t(n-6, 3n/2-5, 2n-3)$ can be sorted with only 2-moves according to one of the basic cases. The permutations resulted by the reduction are one of the basic cases stated in Lemma 4.

Since only the first transposition applied to $u_{n,\ell^*} \uplus u_{n,\ell^*}$ is a 0-move, we have used only one additional transposition than the lower bound. Therefore there is a sequence of $n+1$ transpositions that sorts $u_{n,\ell^*} \uplus u_{n,\ell^*}$. $\square$

Lu and Yang [12] claimed that their computational analysis proved that $u_{10,8}$ is a super-bad permutation, for $u_{10,8} \uplus u_{10,8}$ would require at least 12 transpositions to sort according to them. However, Example 2 provides a sequence of only 11 transpositions that sorts $u_{10,8} \uplus u_{10,8}$. Example 2 uses the results in Lemma 4 to obtain such a sequence, and Theorem 6 guarantees that it is a shortest sequence.

Example 2. The permutation $u_{10,8} \uplus u_{10,8}$ can be sorted applying exactly one 0-move.

By Lemma 4, first we apply $t_1 = t(1,7,12)$, $t_2 = t(2,8,13)$ and $t_3 = t(6,12,17)$, and on $gl((u_{10,8} \uplus u_{10,8})t_1t_2t_3)$ the transpositions $t(1,4,9)$, $t(2,5,15)$, $t(10,14,18)$, $t(5,9,15)$, $t(2,8,17)$, $t(1,7,13)$ in this order, and finally apply $t(5,9,15)$ and $t(2,7,13)$.

Notice that the transpositions t_1, t_2 and t_3 — of which only t_1 is a 0-move — when applied in this order, transform $u_{10,8} \uplus u_{10,8}$ into a permutation that is equivalent by reduction to $[7\,4\,2\,13\,10\,17\,15\,6\,3\,1\,8\,5\,12\,9\,16\,14\,11]$, which can be sorted using just 2-moves.

4   Further Remarks about Unions of Lonely Permutations

The goal of this paper is to establish the nonexistence of super-bad permutations. Nevertheless, we agree that the union of lonely permutations might give us families of permutations with transposition distance close to the diameter. Theorem 6 shows that it is not possible to push the lower bound on the transposition diameter using unions of two copies of a knot. This is a strong evidence that the strategy of using unions of two copies of the same lonely permutation is not good to find diametral permutations.

However, by using Dias's implementation [5] of a branch-and-bound algorithm – modified to take advantage of the toric equivalence classes – to determine exactly the transposition distance, we show that the permutation $u_{4,4} \uplus u_{10,8}$ has transposition distance equal to 9, the same as $u_{4,4} \uplus u_{10,10}$ [7]. The fact that $(u_{10,8})^{-1} = u_{10,7}$ implies that $(u_{4,4} \uplus u_{10,8})^{-1} = (u_{4,4} \uplus u_{10,7})$, and so the permutation $u_{4,4} \uplus u_{10,7}$ has the same transposition distance 9. Therefore, the argument given by [6] can be applied to construct new families $u_{4,4} \uplus u_{10,8} \uplus u_{3,3} \ldots \uplus u_{3,3}$ and $u_{4,4} \uplus u_{10,7} \uplus u_{3,3} \ldots \uplus u_{3,3}$ that meet the current lower bound. Refer to Table 1 to see that the only unions of the lonely permutations $u_{4,4}$ and the possible $u_{10,\ell}$ that require two 0-moves are $u_{4,4} \uplus u_{10,10}$, $u_{4,4} \uplus u_{10,8}$ and $u_{4,4} \uplus u_{10,7}$.

Table 1. Distances for $u_{10,\ell}$ and $u_{4,4} \uplus u_{10,\ell}$. Shaded cells correspond to three families that meet the current lower bound of [6].

π	$u_{10,2}$	$u_{10,3}$	$u_{10,4}$	$u_{10,5}$	$u_{10,6}$	$u_{10,7}$	$u_{10,8}$	$u_{10,9}$	$u_{10,10}$
$d_t(\pi)$	5	6	6	6	5	6	6	6	6
$d_t(u_{4,4} \uplus \pi)$	8	8	8	8	8	9	9	8	9

References

1. Bafna, V., Pevzner, P.A.: Sorting by Transpositions. SIAM J. Disc. Math. 11, 224–240 (1998)
2. Boore, J.L.: The duplication/random loss model for gene rearrangement exemplified by mitochondrial genomes of deuterostome animals. In: Comparative Genomics, pp. 133–148. Kluwer Academic Publishers (2000)
3. Bulteau, L., Fertin, G., Rusu, I.: Sorting by Transpositions Is Difficult. In: Aceto, L., Henzinger, M., Sgall, J. (eds.) ICALP 2011, Part I. LNCS, vol. 6755, pp. 654–665. Springer, Heidelberg (2011)
4. Christie, D.A.: Genome Rerrangement Problems. Ph.D. dissertation, University of Glasgow, Scotland (1999)
5. Dias, Z.: Rearranjo de genomas: uma coletânea de artigos. Ph.D. dissertation, UNICAMP, Brazil (2002)
6. Elias, I., Hartman, T.: A 1.375-approximation algorithm for sorting by transpositions. IEEE/ACM Trans. Comput. Biol. Bioninformatics 3(4), 369–379 (2006)
7. Eriksson, H., Eriksson, K., Karlander, J., Svensson, L., Wästlund, J.: Sorting a bridge hand. Discrete Math. 241(1), 289–300 (2001)
8. Fortuna, V.J.: Distâncias de Transposição entre Genomas. Master dissertation, Instituto de Computação – UNICAMP, Brazil (2005)
9. Hausen, R.A., Faria, L., de Figueiredo, C.M.H., Kowada, L.A.B.: On the Toric Graph as a Tool to Handle the Problem of Sorting by Transpositions. In: Bazzan, A.L.C., Craven, M., Martins, N.F. (eds.) BSB 2008. LNCS (LNBI), vol. 5167, pp. 79–91. Springer, Heidelberg (2008)

10. Hausen, R.A., Faria, L., de Figueiredo, C.M.H., Kowada, L.A.B.: Unitary Toric Classes, the Reality and Desire Diagram, and Sorting by Transpositions. SIAM J. Disc. Math. 24(3), 792–807 (2010)
11. Kowada, L.A.B., Hausen, R.A., de Figueiredo, C.M.H.: Bounds on the Transposition Distance for Lonely Permutations. In: Ferreira, C.E., Miyano, S., Stadler, P.F. (eds.) BSB 2010. LNCS, vol. 6268, pp. 35–46. Springer, Heidelberg (2010)
12. Lu, L., Yang, Y.: A Lower Bound on the Transposition Diameter. SIAM J. Disc. Math. 24(4), 1242–1249 (2010)
13. Meidanis, J., Walter, M.E.M.T., Dias, Z.: Transposition distance between a permutation and its reverse. In: Proceedings of the 4th South American Workshop on String Processing, pp. 70–79. Carleton University Press, Valparaíso (1997)
14. Sankoff, D., Leduc, G., Antoine, N., Paquin, B., Lang, B.F., Cedergren, R.: Gene sort comparisons for phylogenetic inference: evolution of the mitochondrial genome. Proc. Natl. Acad. Sci. 89(14), 6575–6579 (1992)

Extending the Algebraic Formalism for Genome Rearrangements to Include Linear Chromosomes

Pedro Feijao[1] and Joao Meidanis[1,2]

[1] Institute of Computing, University of Campinas
`ra932015@ic.unicamp.br`
[2] Scylla Bioinformatics
`meidanis@scylla.com.br`

Abstract. Algebraic rearrangement theory, as introduced by Meidanis and Dias, focuses on representing the order in which genes appear in chromosomes, and applies to circular chromosomes only. By shifting our attention to genome adjacencies, we are able to extend this theory to linear chromosomes in a very natural way, and extend the distance formula to the general multichromosomal case, with both linear and circular chromosomes. The resulting distance, which we call algebraic distance here, is very similiar to, but not quite the same as, DCJ distance. We present linear time algorithms to compute it and to sort genomes. We also show how to compute the algebraic distance from the adjacency graph. Some results on more general k-break distances are given, with algebraic distance being 2-break distance under our interpretation.

Keywords: genome rearrangement.

1 Introduction

The genome rearrangement problem can be seen as follows. We are given two genomes π and σ and need to return a most parsimonious sequence of rearrangement operations that transforms π into σ. Important issues to consider are the kinds of operations permitted, as well as the notion of parsimony used.

Typically, allowed operations include reversals, transpositions, translocations, fusions, fissions, and, at times, block interchanges. These are what we will term *classical operations*. Some of them have been treated in isolation in the literature; in other articles, two or more operation types were considered. In general, parsimony is defined as using the fewest operations. Sometimes, however, especially when more than one operation kind is involved, different weights are assigned to each kind, and a solution of minimum total weight is sought. Weight choice is still a matter of debate.

As research on this topic progresses, we see models including more and more operation types. There is a balance between including all biologically sound, relevant operations, and the possibility of solving the resulting combinatorial problem efficiently. Yancopoulos et al. [1] introduced a new operation, called *double-cut-and-join* (DCJ), that models all classical operations, with weight 2 for

M.C.P. de Souto and M.G. Kann (Eds.): BSB 2012, LNBI 7409, pp. 13–24, 2012.
© Springer-Verlag Berlin Heidelberg 2012

transpositions and block interchanges, and weight 1 for reversals, translocations, fusions, and fissions. This weight assigment is not unlike many others used in the literature [2–4].

In terms of formal models, the first papers on genome rearrangements dealt with unichromosomal genomes, and represented chromosomes as lists of genes, possibly signed to indicate orientation. This representation is not unique, though, because the reverse complement of a gene list represents the same chromosome. Besides, a circular chromosome can be represented by several lists, depending on where one starts the list.

In 2000, Meidanis and Dias [5] introduced a way of looking at chromosomes as bijections from genes to genes, that is, permutations over a certain gene set. This algebraic formalism is interesting because many results can be stated and proved in the realm of permutation group theory. The approach applies directly to circular chromosomes; for linear chromosomes, the usual tactic is to circularize them by means of "dummy" elements, which are removed at the end. Many papers were written on this subject. Huang and Lu [6] present a relatively recent account.

Another way of uniquely specifying a multichromosomal genome, with both linear and circular chromosomes, is by listing the adjacencies between gene extremities. Bergeron, Mixtacki, and Stoye [7] used this representation and a structure called the adjacency graph, to unify the study of rearrangement problems. Feijao and Meidanis [8] also used it to define a new operation, *single-cut-or-join* (SCJ), which gives rise to a distance related to the number of breakpoints between two genomes, and for which hard problems, such as finding a median genome, have straightforward, efficient solutions.

In this paper, we suggest a new way of modeling genomes which can be seen as a mixture of algebraic rearrangement theory with the adjacency formalism. As a result, linear chromosomes are modeled directly, without the need of "dummy" elements. The new theory inherits many results from the algebraic theory, including the elegant distance formula $d = \frac{\|\sigma\pi^{-1}\|_2}{2}$. Sorting by DCJ operations can be achieved in linear time [7], provided one does not insist in knowing the *type* of each operation, just the operations themselves. Contrast this with the results from Mira and Meidanis [3], and Huang and Lu [6], who output typed operations but take quadratic time.

The weights assigned to the operations are slightly different from the DCJ weights. As a result, the two distances are close but not identical. This opens up the possibility of solving hard problems, such as finding medians, or optimally reconstructing ancestors in a given phylogenetic tree, more efficiently.

2   Algebraic Rearrangement Theory

Meidanis and Dias [5] introduced a model where permutation group theory was used to model genome rearrangement problems, the *Algebraic rearrangement theory*. In its original form, it is limited to circular chromosomes only, and it was used to solve several problems with different rearrangement operations [2–4].

We will present some basic concepts of this theory and then introduce an extended algebraic theory, which we call *adjacency algebraic theory*, that will allow the modeling of linear chromosomes.

2.1 Basic Concepts

Given a set E, a *permutation* α is a map from E onto itself, that is, $\alpha : E \to E$. Permutations are represented with each element followed by its image. For instance, with $E = \{a, b, c\}$, $\alpha = (a\ b\ c)$ is the permutation where a is mapped to b, which is mapped to c, which in turn is mapped back to a. This representation is not unique; $(b\ c\ a)$ and $(c\ a\ b)$ are equivalent. Figure 1(a) gives a graphical representation of this permutation.

Permutations are composed of one or more *cycles*. For instance, the permutation $\alpha = (a\ b\ c)(d\ e)(f)$ has three cycles. A cycle with k elements is called a *k-cycle*. An 1-cycle represents a fixed element in the permutation and is usually omitted.

The *support* of a permutation is the set of its non-fixed elements. In the previous example, $Supp(\alpha) = \{a, b, c, d, e\}$.

The *product* or *composition* of two permutations α, β is denoted by $\alpha\beta$. The product $\alpha\beta$ is defined as $\alpha\beta(x) = \alpha(\beta(x))$ for $x \in E$. For instance, with $E = \{a, b, c, d, e, f\}$, $\alpha = (b\ d\ e)$ and $\beta = (c\ a\ e\ b\ f\ d)$, we have $\alpha\beta = (c\ a\ b\ f\ e\ d)$.

In general $\alpha\beta \neq \beta\alpha$, but when α and β are disjoint cycles, that is, don't have any element in common, they commute: $\alpha\beta = \beta\alpha$. Every permutation can be written in an unique way as a product of disjoint cycles; this is called the *cycle decomposition* of a permutation.

The *identity permutation*, which maps every element into itself, will be denoted by $\mathbf{1}$. Every permutation α has an *inverse* α^{-1} such that $\alpha\alpha^{-1} = \alpha^{-1}\alpha = \mathbf{1}$. For a cycle, the inverse is obtained by reverting the order of its elements: $(c\ b\ a)$ is the inverse of $(a\ b\ c)$.

The *conjugation* of β by α, denoted by $\alpha \cdot \beta$, is the permutation $\alpha\beta\alpha^{-1}$. This results in a permutation with the same structure as β, but with α applied to each element. For instance, if $\beta = (b_1\ b_2\ \ldots\ b_n)$ then $\alpha \cdot \beta = (\alpha b_1\ \alpha b_2\ \ldots\ \alpha b_n)$, where αb_i is a simpler notation for $\alpha(b_i)$.

A *k-cycle decomposition* of a permutation α is a representation of α as a product of k-cycles, not necessarily disjoint. All permutations have a 2-cycle decomposition. The *k-norm* of a permutation α, denoted by $\|\alpha\|_k$, is the minimum number of cycles in a k-cycle decomposition of α. The *norm* of a permutation is defined as the 2-norm, and the subscript can be omitted, that is, $\|\alpha\| \equiv \|\alpha\|_2$.

2.2 Modeling Genomes with Algebraic Theory

We base our definitions on the pioneering work by Meidanis and Dias [5], except that their γ is written as Γ here, following more recent literature [3]. Let $E_n = \{-1, +1, -2, +2, \ldots, -n, +n\}$, where n is the number of genes, as the base set to model genomes as permutations, representing all genes in both orientations.

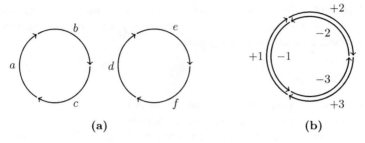

Fig. 1. (a) Graphic representation of the permutation $(a\ b\ c)(d\ e\ f)$, composed by two 3-cycles; (b) a circular genome represented by the permutation $\pi = (+1\ +2\ +3)(-3\ -2\ -1)$, with two 3-cycles, one for each strand

Let Γ be the permutation that maps each gene into its reverse complement, that is, $\Gamma = (-1\ +1)(-2\ +2)\ldots(-n\ +n)$. A *chromosome* is the product of two cycles α and $\Gamma \cdot \alpha^{-1}$, representing the strands of the chromosome. A *genome* is the product of disjoint chromosomes. A necessary and sufficient condition for a permutation π to represent a valid genome is: (1) $\Gamma\pi\Gamma = \pi^{-1}$ and (2) no strand in π contains both $-i$ and $+i$ for any gene i.

For instance, the circular genome depicted in Figure 1 is modelled by $\pi = (+1\ +2\ +3)(-3\ -2\ -1)$. Notice that $(+1\ +2\ +3) = \Gamma \cdot (-3\ -2\ -1)^{-1}$, making it a valid product of cycles representing a chromosome, and also guaranteeing $\pi^{-1} = \Gamma\pi\Gamma$.

A *rearrangement operation* ρ applicable to a genome π is defined as a permutation ρ for which $\pi' = \rho\pi$ is a valid genome. The *weight* of the operation ρ is defined as $\|\rho\|/2$. With this definition, circular fusions and fissions are modelled with permutations formed with two 2-cycles, and have therefore weight 1, whereas transpositions are modelled with two 3-cycles, and block interchanges with four 2-cycles, both with a resulting weight of 2 [3]. This weight definition agrees with the DCJ weights for reversals and generalized transpositions, but differs from it on fusions and fissions, which are weighted 1 in DCJ [1] and 1/2 here.

With this background, we can formulate the *algebraic rearrangement problem* as finding permutations $\rho_1, \rho_2, \ldots, \rho_n$ that minimally transform π into σ, that is, $\rho_i \ldots \rho_2\rho_1\pi$ is a valid genome for every i, $\rho_n \ldots \rho_2\rho_1\pi = \sigma$, and $\sum_{i=1}^{n} \|\rho_i\|/2$ is minimum. This minimum value is called the *algebraic distance* between π and σ, denoted by $d(\pi, \sigma)$.

It is not hard to see that $d(\pi, \sigma) = \|\sigma\pi^{-1}\|/2$: we can easily show that $\|\sigma\pi^{-1}\|/2$ is a lower bound for the distance with some algebraic manipulation of the definition. If $\rho_1, \rho_2, \ldots, \rho_n$ minimally transform π into σ, we have:

$$\frac{\|\sigma\pi^{-1}\|}{2} = \frac{\|\rho_n \ldots \rho_2\rho_1\|}{2} \leq \frac{\sum_{i=1}^{n} \|\rho_i\|}{2} = d(\pi, \sigma) \tag{1}$$

On the other hand, $\sigma\pi^{-1}$ itself can be considered a (possibly very heavy) rearrangement operation, because $(\sigma\pi^{-1})\pi = \sigma$ is a valid genome, and its weight equals the lower bound, showing that this is in fact the distance value.

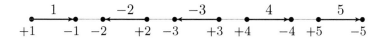

Fig. 2. A genome composed of just a linear chromosome, with gene extremities labeled according to the set theoretical genome representation, also used in our adjacency algebraic theory

Classical definition restricts the available operations to small weight permutations, such as just fusions, fissions, reversals, etc. We will show in Section 3.2 that one can achieve the same distance using only operations of weight 1 or less.

Although several results in genome rearrangement theory have been achieved using this model, in its original form it can only model circular chromosomes. We will expand it to allow linear chromosomes as well by introducing an additional genome representation scheme, which will expand the theory but at the same time maintain most of its important results.

2.3 Adjacency Algebraic Theory

The original algebraic formulation focuses on representing the order in which genes appear in chromosomes, and the "circular" nature of permutations forces it to be applied to circular chromosomes only. In our proposed formulation, we shift our attention to genome adjacencies, adequately extending this theory to the general multichromosomal case, with both linear and circular chromosomes, while keeping most of the properties of the original formulation. We call it the *adjacency algebraic theory*.

This formulation is similar to the set representation of a genome [7, 8]. In this representation, each gene a has two *extremities*, called *tail* and *head*, respectively denoted by a_t and a_h, or alternatively using signs, where $-a = a_h$ and $+a = a_t$. An *adjacency* is an unordered pair of extremities indicating a linkage between two consecutive genes in a chromosome. An extremity not adjacent to any other extremity in a genome is called a *telomere*. A genome is represented as a set of adjacencies and telomeres (possibly omitted, when the gene set is given) where each extremity appears at most once. For instance, the genome in Fig. 2 is represented by $\{\{+1\}, \{-1, -2\}, \{+2, -3\}, \{+3, +4\}, \{-4, +5\}, \{-5\}\}$ or just by $\{\{-1, -2\}, \{+2, -3\}, \{+3, +4\}, \{-4, +5\}\}$.

In the adjacency algebraic theory, genomes are also representated by permutations, but a *genome* is a product of 2-cycles, where each 2-cycle corresponds to an adjacency. Therefore, the genome in Fig. 2 is represented as $\pi' = (-1 \ -2)(+2 \ -3)(+3 \ +4)(-4 \ +5)$. Note that in this representation telomeres can be safely omitted, since they are 1-cycles. With this formulation, linear chromosomes can be represented.

The first property of this new representation is that there is a direct relationship with the original one. Specifically, if π is a permutation representing a circular genome in the original algebraic formulation, then $\pi' = \pi\Gamma$ represents the same genome in the adjacency theory. For instance, for the genome in Fig. 1, we have $\pi = (+1 +2 +3)(-3 -2 -1)$ and $\pi' = \pi\Gamma = (-1 +2)(-2 +3)(-3 +1)$.

Another property is that the rearrangement events are represented by the same permutations in both formulations. If π and σ are genomes in the original formulation and $\rho\pi = \sigma$, then by multiplying by Γ on the right we easily get $\rho\pi' = \sigma'$, where π' and σ' represent the same genomes in the adjacency formulation. Also, $\sigma'\pi'^{-1} = \sigma\Gamma(\pi\Gamma)^{-1} = \sigma\Gamma\Gamma^{-1}\pi^{-1} = \sigma\pi^{-1}$. So, both the rearrangement operations and the all-important permutation $\sigma\pi^{-1}$ remain the same in the adjacency theory, meaning that we have many results (all related to circular genomes) already demonstrated. Therefore, we will have to develop new results in order to extend the algebraic theory for linear chromosomes, specifically to treat the cases where telomeres appear in the permutations, as we will see in the following sections.

3   Sorting by Algebraic Operations

A permutation ρ is called a *sorting operation* from π to σ if $\rho\pi$ is a valid genome and $d(\rho\pi, \sigma) = d(\pi, \sigma) - \|\rho\|/2$, that is, applying ρ in π decreases the distance between π and σ by the weight of operation ρ, that we already defined as being half the norm. But we see that $d(\rho\pi, \sigma) = d(\pi, \sigma) - \|\rho\|/2 \Rightarrow \|\sigma(\rho\pi)^{-1}\|/2 = \|\sigma\pi^{-1}\|/2 - \|\rho\|/2 \Rightarrow \|(\sigma\pi^{-1})\rho^{-1}\| = \|\sigma\pi^{-1}\| - \|\rho\|$. Therefore, to say that ρ decreases the distance by its weight is the same as $\rho|\sigma\pi^{-1}$, that is, ρ divides $\sigma\pi^{-1}$. Then, a sorting operation from π to σ is a valid operation on π that divides $\sigma\pi^{-1}$. Some results on valid operations and divisibility are given in the following lemmas:

Lemma 1 (Valid operations). *Given a genome π, a permutation ρ is a valid operation on π, that is, $\rho\pi$ is a valid genome, if and only if $\pi \cdot \rho = \rho^{-1}$.*

Proof. A permutation π is a valid genome if $\pi^2 = 1$. If $\rho\pi$ is valid, then $(\rho\pi)^2 = 1$ and $(\rho\pi)^2 = \rho\pi\rho\pi = \rho(\pi \cdot \rho) = 1$, since $\pi = \pi^{-1}$. Then $\pi \cdot \rho$ is the inverse of ρ. On the other hand, if $\pi \cdot \rho = \rho^{-1}$, then $(\rho\pi)^2 = \rho\pi\rho\pi = \rho(\pi \cdot \rho) = \rho\rho^{-1} = 1$. □

Corollary 1. *Any permutation ρ that can be written as $\rho = \mu(\pi \cdot \mu^{-1})$ is a valid operation in π.*

Proof. It is easy to see that $\pi \cdot \rho = \rho^{-1}$, then by Lemma 1 ρ is valid on π. □

This corollary is important because permutations in the form $\rho = \alpha(\pi \cdot \alpha^{-1})$ will be the basis of our sorting operations, as we will see when we study the permutation $\sigma\pi^{-1}$ below. Now we will show an important result about permutation divisibility, that applying the reverse conjugation mantains divisibility:

Lemma 2. *Given a genome π and a permutation α, for any permutation μ where $\mu|\alpha$ we have $\pi \cdot \mu^{-1}|\pi \cdot \alpha^{-1}$.*

Proof. Since $\mu|\alpha$, we have $\|\alpha\mu^{-1}\| = \|\alpha\| - \|\mu\|$. Then:

$$\|(\pi \cdot \alpha^{-1})(\pi \cdot \mu^{-1})^{-1}\| = \|\pi \cdot (\alpha^{-1}\mu)\| = \|\alpha^{-1}\mu\| = \|\alpha\mu^{-1}\| = \|\alpha\| - \|\mu\|$$

Since the operations of conjugation and inverse do not change the norm of a permutation, we have $\|\alpha\| = \|\pi \cdot \alpha^{-1}\|$ and $\|\mu\| = \|\pi \cdot \mu^{-1}\|$, and then

$$\|(\pi \cdot \alpha^{-1})(\pi \cdot \mu^{-1})^{-1}\| = \|\pi \cdot \alpha^{-1}\| - \|\pi \cdot \mu^{-1}\|$$

the exact definition of $\pi \cdot \mu^{-1} | \pi \cdot \alpha^{-1}$. $\square$

Using Lemma 2 we get to the following lemma, where the proof was left out for space reasons.

Lemma 3. *Given a genome π and a permutation $\tau = \alpha(\pi \cdot \alpha^{-1})$ where $\|\tau\| = \|\alpha\| + \|\pi \cdot \alpha^{-1}\|$, for any permutation μ where $\mu | \alpha$ we have that the permutation $\rho = \mu(\pi \cdot \mu^{-1})$ divides τ, that is, $\rho | \tau$, and also ρ is a valid operation on π.*

It should be noted that finding a cycle $\mu = (e_1 \ \ldots \ e_k)$ that divides a cycle α is easy, if we choose $e_1, \ldots, e_k$ as a subset of elements of α such that they also appear in α in the same order. For instance, if $\alpha = (1\ 2\ 3\ 4\ 5)$, then $\mu = (1\ 3\ 4)$ divides α, but $\mu' = (1\ 4\ 3)$ does not. A formal proof of this was shown by Huang and Lu [6, Lemma 2.7].

A last lemma exposes the relationship of these $\mu(\pi \cdot \mu^{-1})$ operations with the k-break operation, introduced by Alekseyev and Pevzner [9]. A k-break is an operation that cuts k adjacencies in π and then joins k new ones within the same extremities.

Lemma 4. *A permutation $\rho = \mu(\pi \cdot \mu^{-1})$ where μ and $\pi \cdot \mu^{-1}$ are disjoint and $\|\mu\| = k - 1$, is a k-break operation on π.*

Proof. Let $\mu = (e_1 \ e_2 \ \cdots \ e_k)$, so $\|\mu\| = k - 1$, and let $\rho = \mu(\pi \cdot \mu^{-1})$. If μ and $\pi \cdot \mu^{-1}$ are disjoint, π has the adjacencies $(e_1 \ \pi e_1) \ldots (e_k \ \pi e_k)$ and applying ρ we can see that $\rho\pi$ will have the adjacencies $(e_1 \ \pi e_k)(e_2 \ \pi e_1) \ldots (e_k \ \pi e_{k-1})$. Therefore, k adjacencies of π were removed and changed by k new ones, and ρ is a k-break. $\square$

In the next section we will use Lemma 3 to find sorting operations, that is, valid operations that divide $\sigma\pi^{-1}$.

3.1 Characterizing $\sigma\pi^{-1}$ and Finding Sorting Operations

In this section we will use the *Adjacency Graph* between two genomes π and σ, defined by Bergeron, Mixtacki, and Stoye [7]. In this graph, denoted as $AG(\pi, \sigma)$, the vertices are the adjacencies and telomeres of π and σ, and for each $u \in \pi$ and $v \in \sigma$ there is an edge between u and v for each extremity that u and v have in common. We will show that every connected component in $AG(\pi, \sigma)$ has a direct relationship with a permutation in $\sigma\pi^{-1}$ and then determine sorting operations on these permutations.

Cycles in $AG(\pi,\sigma)$. A cycle of size n in $AG(\pi,\sigma)$ contains the following adjacencies as vertices, starting with an adjacency (e_1,e_2) in π and alternating vertices of π and σ:

$$\underbrace{(e_1,e_2)}_{\pi},\underbrace{(e_2,e_3)}_{\sigma},\ldots,\underbrace{(e_{2k-1},e_{2k})}_{\pi},\underbrace{(e_{2k},e_{2k+1})}_{\sigma},\ldots,\underbrace{(e_{n-1},e_n)}_{\pi},\underbrace{(e_n,e_1)}_{\sigma}$$

Therefore, adjacencies in π will have the form $\pi_k = (e_{2k-1},e_{2k})$ and in σ the form $\sigma_k = (e_{2k},e_{2k+1})$, for $k = 1,\ldots,n$ (assuming for simplicity that $e_{n+1} \equiv e_1$).

When we calculate the product $\sigma\pi^{-1}$ the adjacencies in the $AG(\pi,\sigma)$ cycle will be 2-cycles disjoint from the rest of the adjacencies in π and σ. Therefore, it will be part of the cycle decomposition of $\sigma\pi^{-1}$. We can multiply the 2-cycles in this $AG(\pi,\sigma)$ cycle to obtain the restriction τ of $\sigma\pi^{-1}$ to the $AG(\pi,\sigma)$ cycle (notice that $\pi_i^{-1} = \pi_i$ for each i):

$$\tau = \sigma_1\sigma_2\ldots\sigma_n\pi_1\ldots\pi_n$$
$$\tau = (e_2\ e_3)\ldots(e_{2k}\ e_{2k+1})\ldots(e_n\ e_1)(e_1\ e_2)\ldots(e_{2k-1}\ e_{2k})\ldots(e_{n-1}\ e_n)$$
$$\tau = (e_n\ e_{n-2}\ \ldots\ e_4\ e_2)(e_1\ e_3\ \ldots\ e_{n-3}\ e_{n-1})$$
$$\tau = (e_n\ e_{n-2}\ \ldots\ e_4\ e_2)(\pi e_2\ \pi e_4\ \ldots\ \pi e_{n-2}\ \pi e_n)$$
$$\tau = \alpha(\pi\cdot\alpha^{-1}),$$

where $\alpha = (e_n\ e_{n-2}\ \ldots\ e_4\ e_2)$. Therefore, a cycle of length n in $AG(\pi,\sigma)$ corresponds to 2 $(n/2)$-cycles in $\sigma\pi^{-1}$, where one is the reversed π-conjugation of the other.

To extract sorting operations in this case, we see that τ satisfies Lemma 3, therefore any μ that divides α generates a sorting operation $\rho = \mu(\pi\cdot\mu^{-1})$, with weight $w = \|\rho\|/2 = \|\mu\|$.

Odd Paths in $AG(\pi,\sigma)$. An odd path of size n in $AG(\pi,\sigma)$, that is, a path with n edges where n is odd, starting with a telomere e_1 in π and ending at a telomere e_n in σ, has vertices

$$\underbrace{(e_1)}_{\pi},\underbrace{(e_1,e_2)}_{\sigma},\underbrace{(e_2,e_3)}_{\pi},\ldots,\underbrace{(e_{2k-1},e_{2k})}_{\pi},\underbrace{(e_{2k},e_{2k+1})}_{\sigma},\ldots,\underbrace{(e_{n-1},e_n)}_{\pi},\underbrace{(e_n)}_{\sigma}$$

Then, similarly to the previous case, computing the restriction τ of $\sigma\pi^{-1}$ to these adjacencies, we have

$$\tau = (\ e_n\ e_{n-2}\ \ldots\ e_3\ e_1\ e_2\ e_4\ \ldots\ e_{n-3}\ e_{n-1})$$

Therefore, an odd path in $AG(\pi,\sigma)$ corresponds to an n-cycle in $\sigma\pi^{-1}$. Notice that we can write this as a product of (nondisjoint) reversed π-conjugates:

$$\tau = (e_n\ e_{n-2}\ \ldots\ e_3\ e_1)(e_1\ e_2\ e_4\ \ldots\ e_{n-3}\ e_{n-1})$$
$$\tau = (e_n\ e_{n-2}\ \ldots\ e_3\ e_1)(\pi e_1\ \pi e_3\ \pi e_5\ \ldots\ \pi e_{n-2}\ \pi e_n) = \alpha(\pi\cdot\alpha^{-1})$$

where $\alpha = (e_n\ e_{n-2}\ \ldots\ e_3\ e_1)$, then τ satisfies Lemma 3, and with any μ such that $\mu|\alpha$ we derive a sorting operation $\rho = \mu(\pi\cdot\mu^{-1})$, with weight $w = \|\rho\|/2 = \|\mu\|$.

Even Paths in $AG(\pi, \sigma)$. An even path of size n in $AG(\pi, \sigma)$ will have both path extremities (telomeres) in the same genome. Then, we have two cases: both telomeres in π or in σ.

i) Both telomeres in σ. If both telomeres are in σ, the vertices are of the form

$$\underbrace{(e_1)}_{\sigma}, \underbrace{(e_1, e_2)}_{\pi}, \underbrace{(e_2, e_3)}_{\sigma}, \ldots, \underbrace{(e_{2k-1}, e_{2k})}_{\pi}, \underbrace{(e_{2k}, e_{2k+1})}_{\sigma}, \ldots, \underbrace{(e_{n-1}, e_n)}_{\pi}, \underbrace{(e_n)}_{\sigma}$$

Then, computing the restriction τ of $\sigma\pi^{-1}$, we have

$$\tau = (e_n \ e_{n-2} \ \cdots \ e_4 \ e_2 \ \pi e_2 \ \pi e_4 \ \cdots \ \pi e_{n-2} \ \pi e_n)$$

which is an n-cycle in $\sigma\pi^{-1}$, and with some manipulation we get to

$$\tau = (\pi e_2 \ e_n)(e_n \ e_{n-2} \ \cdots \ e_4 \ e_2)(\pi e_2 \ \pi e_4 \ \cdots \ \pi e_{n-2} \ \pi e_n)$$
$$\tau = (e_1 \ e_n)\alpha(\pi \cdot \alpha^{-1})$$

that is, the product of a 2-cycle with $\alpha(\pi \cdot \alpha^{-1})$ permutation, where $\alpha = (e_n \ e_{n-2} \ \cdots \ e_4 \ e_2)$.

ii) Both telomeres in π. If both telomeres are in π, the vertices are of the form

$$\underbrace{(e_1)}_{\pi}, \underbrace{(e_1, e_2)}_{\sigma}, \underbrace{(e_2, e_3)}_{\pi}, \ldots, \underbrace{(e_{2k-1}, e_{2k})}_{\sigma}, \underbrace{(e_{2k}, e_{2k+1})}_{\pi}, \ldots, \underbrace{(e_{n-1}, e_n)}_{\sigma}, \underbrace{(e_n)}_{\pi}$$

Then, the restriction τ of $\sigma\pi^{-1}$ will be

$$\tau = (e_{n-1} \ e_{n-3} \ \cdots \ e_3 \ e_1 \ \pi e_3 \ \cdots \ \pi e_{n-3} \ \pi e_{n-1} \ e_n)$$

again an n-cycle in $\sigma\pi^{-1}$. With more algebrism with get to

$$\tau = (e_n \ e_{n-1})(e_{n-1} \ e_{n-3} \ \cdots \ e_3 \ e_1 \ \pi e_3 \ \cdots \ \pi e_{n-3} \ \pi e_{n-1})$$
$$\tau = (e_n \ e_{n-1})\alpha(\pi \cdot \alpha^{-1})$$

and again we get to a product of a 2-cycle with a permutation in the form $\alpha(\pi \cdot \alpha^{-1})$.

In both even path cases, permutation $\alpha(\pi \cdot \alpha^{-1})$ is a part of τ and $\rho = \mu(\pi \cdot \mu^{-1})$, where $\mu|\alpha$ is a sorting operation with weight $w = \|\rho\|/2 = \|\mu\|$.

iii) Special cases. In both types of even paths there is one special case, specifically when $n = 2$, where permutation α becomes an 1-cycle and τ is reduced to only the 2-cycle, and in both cases $\tau = (e_1 \ e_2)$. In the first case, both telomeres are in σ and τ is a *cut* in π, splitting the adjacency of $(e_1 \ e_2)$ of π into two telomeres. It is easy to see that τ is a sorting operation. On the second case, both telomeres are in π, and τ is a *join* in π, joining telomeres e_1 and e_2 into one adjacency. Again, τ is a sorting operation. In both cases, this operation has weight $1/2$, since it is formed by just one 2-cycle.

In this section we learned how to derive sorting operations from the permutation $\sigma\pi^{-1}$. Sorting operations are usually in the format $\rho = \mu(\pi \cdot \mu^{-1})$, where μ divides a cycle of $\sigma\pi^{-1}$, but there are also special cases of single 2-cycle operations like cuts and joins.

3.2   Algebraic Sorting with 2-Break (DCJ) Operations

From the previous section we saw that we can always find a sequence $\rho_1, \ldots, \rho_n$ of sorting operations such that $\rho_n \ldots \rho_1 \pi = \sigma$, and $\sum_{i=1}^{n} \|\rho_i\|/2 = \|\sigma\pi^{-1}\|/2 = d(\pi, \sigma)$. But that leaves the following question: can we always find sorting operations $\rho_1, \ldots, \rho_n$ where $\rho_n \ldots \rho_1 \pi = \sigma$ and $\sum_{i=1}^{n} \|\rho_i\|/2 = d(\pi, \sigma)$, with the additional constraint that $\|\rho_i\|/2 \leq w$, for $i = 1, \ldots, n$, for any given w? It should be noted that when we choose different values of w, the distance does not change, since the weight of the rearrangement operations is always the same, but we change the *scenario* of the rearrangent sorting.

Of particular interest are operations of weight 1 or less, corresponding to 2-breaks, that we know from DCJ theory that correspond to all classic operations of reversal, translocation, fusion and fissons (generalized transpositions are also possible by applying two specific operations of weight 1). From Section 3.1 we see that it is always possible to find 2-break sorting operations. In all cases where the sorting operation is in the format $\rho = \mu(\pi \cdot \mu^{-1})$, if we choose μ as a 2-cycle, then the weight is $\|\rho\|/2 = \|\mu\| = 1$ and ρ is a 2-break. There are also the special cases of cuts or joins, but in both cases the operation has weight $1/2$. Therefore, using algebraic theory, it is always possible to find a rearrangement scenario using only operations of weight 1 or less, which means only classical operations are being used.

Also, it is not difficult to see that the operations found by the algorithm for sorting with DCJ operations by Bergeron, Mixtacki, and Stoye [7] are also sorting operations under the algebraic theory, which means that algebraic sorting by 2-breaks can be achieved in linear time.

3.3   Comparing the Algebraic with DCJ Distance

To compare the algebraic and DCJ distances, we will use the graph $AG(\pi, \sigma)$ again. From Section 3.1, we know that any cycle of size $2n$ in $AG(\pi, \sigma)$ will correspond to two n-cycles in $\sigma\pi^{-1}$, and a path of size n in $AG(\pi, \sigma)$ will become an n-cycle in $\sigma\pi^{-1}$. Since the norm of an n-cycle is $n-1$ and the algebraic weight of an operation is the norm divided by two, the cost of sorting a cycle of size $2n$ is $n-1$, and sorting a path of size n costs $(n-1)/2$. Then, the algebraic distance can be computed as follows:

$$d(\pi, \sigma) = \sum_{k=1}^{n}(k-1)C_{2k} + \sum_{k=1}^{n}\frac{k-1}{2}P_k \qquad (2)$$

where C_{2k} is the number of cycles of size $2k$ and P_k is the number of paths of size k in $AG(\pi, \sigma)$. Also, we know that there are $4N$ extremities in the vertices of $AG(\pi, \sigma)$, where N is the number of genes. Since each cycle of size $2k$ has $2k$ vertices, comprising of $4k$ extremities, and each path of size k has $k+1$ vertices with a total of $2k$ extremities, we have

$$N = \sum_{k=1}^{n}kC_{2k} + \sum_{k=1}^{n}\frac{k}{2}P_k \qquad (3)$$

Using (2) and (3) we have

$$d(\pi,\sigma) = N - \sum_{k=1}^{n} C_{2k} - \sum_{k=1}^{n} \frac{1}{2} P_k = N - (C + \frac{P}{2}) \qquad (4)$$

where $C = \sum_{k=1}^{n} C_{2k}$ and $P = \sum_{k=1}^{n} P_k$ are respectively the number of cycles and paths in $AG(\pi,\sigma)$. Since $d_{DCJ} = N - (C + P_{odd}/2)$ [7], we have

$$d(\pi,\sigma) = d_{DCJ}(\pi,\sigma) - \frac{P_{even}}{2} \qquad (5)$$

where P_{odd} and P_{even} denote the number of odd and even paths in $AG(\pi,\sigma)$, respectively.

This small difference is due to the fact that although most DCJ operations have the same weight in the algebraic theory, when sorting an even path at least one *cut* or *join* must be performed. Since these operations are modelled as permutations with a single 2-cycle, they have weight $1/2$ under the algebraic theory, but weight 1 in the DCJ model, hence the difference in the distances.

Another difference is that the DCJ model allows operations that recombine two even paths into two odd paths [10]. We can see in the distance equations (4) and (5) that this kind of operation is indeed optimal in the DCJ model (that is, it reduces the distance by 1) but in the algebraic model the distance is not changed.

In addition to the distance formula, we also compared algebraic and DCJ distances with a scatterplot between randomly evolved genomes. Starting with a genome with 1000 genes and 5 chromosomes, we applied a random number of rearrangement operations and then measured the distance between the original and evolved genomes under both algebraic and DJC distances, resulting in the scatterplot shown in Figure 3.

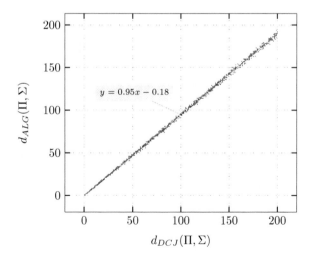

Fig. 3. Scatterplot comparing algebraic and DCJ distances between randomly evolved genomes. Linear regression on the points resulted in the equation $y = 0.95x - 0.18$.

4 Conclusions

The main accomplishment reported in this paper is an extension of the algebraic theory of genome rearrangements to include linear genomes in a quite natural way. The result is a new notion of genomic distance, which can be computed in linear time, and for which a list of small-weight, sorting operations can also be retrieved in linear time. This is better than the quadratic time obtained in previous papers.

In addition, the new distance yields values very close to the acclaimed DCJ distance. The difference is due to the weight given by cuts and joins in the two approaches: DCJ assigns them weight 1, whereas the algebraic theory assigns them weight $1/2$. Since the distances are not exactly the same, the NP-hardness proofs for some important problems under DCJ, such as finding a median, do not immediately apply to the algebraic distance, leaving open the possibility of a polynomial algorithm for these problems. We remark that the recently introduced SCJ distance did just that with respect to breakpoints: allowed linear solutions for NP-hard problems under other, similar BP-like distances [8].

References

1. Yancopoulos, S., Attie, O., Friedberg, R.: Efficient sorting of genomic permutations by translocation, inversion and block interchange. Bioinformatics 21(16), 3340–3346 (2005)
2. Lu, C.L., Huang, Y.L., Wang, T.C., Chiu, H.T.: Analysis of circular genome rearrangement by fusions, fissions and block-interchanges. BMC Bioinformatics 7, 295 (2006)
3. Mira, C., Meidanis, J.: Sorting by block-interchanges and signed reversals. In: ITNG 2007, pp. 670–676 (2007)
4. Dias, Z., Meidanis, J.: Genome rearrangements distance by fusion, fission, and transposition is easy. In: Proc. SPIRE 2001, pp. 250–253 (2001)
5. Meidanis, J., Dias, Z.: An Alternative Algebraic Formalism for Genome Rearrangements. In: Sankoff, D., Nadeau, J.H. (eds.) Comparative Genomics: Empirical and Analyitical Approaches to Gene Order Dynamics, Map Alignment and Evolution of Gene Families, pp. 213–223. Kluwer Academic Publishers (2000)
6. Huang, Y.L., Lu, C.L.: Sorting by reversals, generalized transpositions, and translocations using permutation groups. Journal of Computational Biology 17(5), 685–705 (2010)
7. Bergeron, A., Mixtacki, J., Stoye, J.: A Unifying View of Genome Rearrangements. In: Bücher, P., Moret, B.M.E. (eds.) WABI 2006. LNCS (LNBI), vol. 4175, pp. 163–173. Springer, Heidelberg (2006)
8. Feijão, P., Meidanis, J.: SCJ: a breakpoint-like distance that simplifies several rearrangement problems. IEEE/ACM Transactions on Computational Biology and Bioinformatics 8, 1318–1329 (2011)
9. Alekseyev, M., Pevzner, P.: Multi-break rearrangements and chromosomal evolution. Theoretical Computer Science 395(2-3), 193–202 (2008)
10. Braga, M.D.V., Stoye, J.: The solution space of sorting by DCJ. Journal of Computational Biology 17(9), 1145–1165 (2010)

On the Approximation Ratio
of Algorithms for Sorting by Transpositions
without Using Cycle Graphs

Gustavo Rodrigues Galvão and Zanoni Dias

University of Campinas, Institute of Computing, Brazil
gustavo.galvao@students.ic.unicamp.br, zanoni@ic.unicamp.br

Abstract. We study the problem of sorting by transpositions, which is related to comparative genomics. Our goal is to determine how good approximation algorithms which do not rely on the cycle graph are when it comes to approximation ratios by implementing three such algorithms. We compare their theoretical approximation ratio to the experimental results obtained by running them for all permutations of up to 13 elements. Our results suggest that the approaches adopted by these algorithms are not promising alternatives in the design of approximation algorithms with low approximation ratios. Furthermore, we prove an approximation bound of 3 for a constrained version of one algorithm, and close a missing gap on the proof for the approximation ratio of another algorithm.

1 Introduction

A transposition is the rearrangement event that switches the location of two contiguous portions of a genome. The problem of computing the transposition distance between two genomes consists in finding the minimum number of transpositions needed to transform one genome into another, and it is usually stated as the Problem of Sorting by Transpositions. Such problem arises in comparative genomics because the transposition distance can be used to estimate the evolutionary distance between two genomes.

The Problem of Sorting by Transpositions was recently proven to be NP-hard [3], therefore it is not likely that a polynomial-time algorithm exists for this problem. It was introduced by Bafna and Pevzner [1], who presented a 1.5-approximation algorithm for it that runs in quadratic time. Later, Elias and Hartman [5] improved the approximation bound to 1.375 (maintaining quadratic time complexity), and this has been the best known algorithmic result. For a more complete reference to the Problem of Sorting by Transpositions, the reader is referred to [6].

Both approximation algorithms cited previously are based on a structure named cycle graph. In an attempt to bypass this structure, some researches posed new approaches to tackle the problem: Benoît-Gagné and Hamel [2] developed a 3-approximation algorithm based on permutation codes that runs in $O(n^2)$ time; Walter, Dias, and Meidanis [10] presented a 2.25-approximation algorithm based on a structure named breakpoint diagram that also runs in $O(n^2)$

M.C.P. de Souto and M.G. Kann (Eds.): BSB 2012, LNBI 7409, pp. 25–36, 2012.

time; and Guyer, Heath, and Vergara [9] devised a heuristic based on the longest increasing subsequence of a permutation that runs in $O(n^5\log n)$ time. In this work we implement these three algorithms, and measure their performance in practice using GRAAu [7]. Although a similar test was ran for each algorithm by its authors, our results bring new insights into the approximation ratios of these algorithms.

While Benoît-Gagné and Hamel [2] concluded from their experiments that the approximation ratio of their 3-approximation could be lowered, our experimental results indicate that it may not. Furthermore, we show that previous experimental data on the approximation factor of the 2.25-approximation algorithm are incorrect, and then we present new experimental results suggesting that its approximation ratio may be lowered to 2. Finally, we prove an approximation bound of 3 for a constrained version of Guyer, Heath, and Vergara's heuristic [9], and the experimental data we obtained suggest that this bound cannot be lower than 2.25.

The remainder of this paper is organized as follows. In the next section, we give the definitions and notation of the paper. Section 3 briefly describes the approximation algorithms studied in this paper, closes a missing gap on Benoît-Gagné and Hamel's proof [2] for the approximation ratio of their algorithm (Lemma 4), and demonstrates that a constrained version of Guyer, Heath, and Vergara's algorithm [9] still has an approximation ratio of 3 (Lemma 5 and Theorem 1). In Sect. 4, we present the experimental results along with a discussion on the approximation ratio of the algorithms. The last section concludes the paper.

2 Preliminaries

We represent genomes as permutations, where genes appear as elements. A *permutation* π is a bijection of $\{1, 2, \ldots, n\}$ onto itself. The group of all permutations of $\{1, 2, \ldots, n\}$ is denoted by S_n, and we write a permutation $\pi \in S_n$ as $\pi = (\pi_1 \ \pi_2 \ \ldots \ \pi_n)$.

A *transposition* is an operation $\rho(i, j, k)$, $1 \leq i < j < k \leq n+1$, that moves blocks of contiguous elements of a permutation $\pi \in S_n$ in such way that $\rho(i, j, k) \cdot (\pi_1 \ \ldots \ \pi_{i-1} \ \underline{\pi_i \ \ldots \ \pi_{j-1}} \ \pi_j \ \ldots \ \pi_{k-1} \ \pi_k \ \ldots \ \pi_n) = (\pi_1 \ \ldots \ \pi_{i-1} \ \underline{\pi_j \ \ldots \ \pi_{k-1}} \ \pi_i \ \ldots \ \pi_{j-1} \ \pi_k \ \ldots \ \pi_n)$. The Problem of Sorting by Transpositions consists in finding the minimum number of transpositions that transform a permutation $\pi \in S_n$ into the identity permutation $I_n = (1 \ 2 \ \ldots \ n)$. This number is known as the *transposition distance* of permutation π and it is denoted by $d(\pi)$.

Given a permutation $\pi \in S_n$, the *left and right codes of an element* π_i, denoted $lc(\pi_i)$ and $rc(\pi_i)$ respectively, are defined as $lc(\pi_i) = |\{\pi_j : \pi_j > \pi_i$ and $1 \leq j \leq i-1\}|$ and $rc(\pi_i) = |\{\pi_j : \pi_j < \pi_i$ and $i+1 \leq j \leq n\}|$. The *left (resp. right) code of a permutation* π is then defined as the sequence of lc's (resp. rc's) of its elements, and it is denoted by $lc(\pi)$ (resp. $rc(\pi)$).

Let us call *plateau* any maximal length sequence of contiguous elements in a number sequence that have the same nonzero value. The number of plateaux in a code c is denoted p(c). We denote by $p(\pi)$ the minimum of $p(lc(\pi))$ and $p(rc(\pi))$. Note that I_n is the only permutation in S_n having zero plateaux.

Example 1. Let $\pi = (5\ 3\ 2\ 4\ 1)$. We have that $lc(\pi) = lc(\pi_1)\ lc(\pi_2)\ lc(\pi_3)\ lc(\pi_4)$
$lc(\pi_5) = 0\ 1\ 2\ 1\ 4$, and $rc(\pi) = rc(\pi_1)\ rc(\pi_2)\ rc(\pi_3)\ rc(\pi_4)\ rc(\pi_5) = 4\ 2\ 1\ 1\ 0$.
Then, $p(\pi) = \min\{p(lc(\pi)), p(rc(\pi))\} = \min\{4, 3\} = 3$.

Given a permutation $\pi \in S_n$, we extend it with two elements $\pi_0 = 0$ and π_{n+1}
$= n + 1$. The extended permutation is still denoted π. A *breakpoint* of $\pi \in S_n$
is a pair of adjacent elements that are not consecutive, that is, a pair (π_i, π_{i+1})
such that $\pi_{i+1} - \pi_i \neq 1$, $0 \leq i \leq n$. The number of breakpoints of π is denoted
by $b(\pi)$. Note that I_n is the only permutation in S_n having zero breakpoints.
Since a transposition can remove at most three breakpoints, we can state the
following lemma.

Lemma 1. *[1] For any permutation $\pi \in S_n$, $d(\pi) \geq \frac{b(\pi)}{3}$.*

A *strip* of a permutation π is a maximal series of consecutive elements without
a breakpoint. We denote the number of strips in π by $s(\pi)$.

Example 2. Let $\pi = (0\ 4\ 5\ 2\ 3\ 1\ 6)$ be the extend permutation of $(4\ 5\ 2\ 3\ 1)$.
We have that the pairs $(0, 4)$, $(5, 2)$, $(3, 1)$, and $(1, 6)$ are breakpoints, thus $b(\pi)$
$= 4$. We also have that $4\ 5$, $2\ 3$, and 1 are strips of π, thus $s(\pi) = 3$.

Let $\pi \in S_n$, $\pi \neq I_n$, s_1 be the first strip of π, and s_m be the last strip of π, m
$\leq n$. If we assume that $\pi_1 = 1$, we can reduce π to a permutation $\sigma \in S_{n-|s_1|}$
such that $\sigma_i = \pi_i - |s_1|$, $i > |s_1|$. It is not hard to see that $d(\pi) = d(\sigma)$. An
analogous argument can be used to show that we can reduce π to a permutation
$\gamma \in S_{n-|s_m|}$ such that $d(\pi) = d(\gamma)$ if $\pi_n = n$. We call *irreducible* any permutation
in which such reductions cannot be applied, and we denote by S_n^* the set formed
by all irreducible permutations of S_n.

Lemma 2. *For any permutation $\pi \in S_n^*$, $s(\pi) = b(\pi) - 1$.*

Proof. Let s_1, s_2, ..., $s_{s(\pi)}$ be the strips of a permutation $\pi \in S_n^*$. The last
element of strip s_i and the first element of strip s_{i+1}, $1 \leq i \leq s(\pi) - 1$, form a
breakpoint, and the pairs (π_0, π_1) and (π_n, π_{n+1}) are always breakpoints on an
irreducible permutation. Therefore, we have that $b(\pi) = s(\pi) + 1$ and the claim
follows.   □

An *increasing subsequence* of a permutation π is a subsequence $\pi_{i_1}\ \pi_{i_2} \ldots \pi_{i_j}$ of
nonnecessarily contiguous elements of π such that for all k, $0 < k < j$, we have
$i_k < i_{k+1}$ and $\pi_{i_k} < \pi_{i_{k+1}}$. A *longest increasing subsequence* is an increasing
subsequence of π of maximum length. The set of the elements belonging to a
longest increasing subsequence is denoted by $\mathrm{LIS}(\pi)$. It is easy to see that, for
any $\pi \in S_n$, $|\mathrm{LIS}(\pi)| = n$ if and only if $\pi = I_n$.

Example 3. Let $\pi = (2\ 3\ 1\ 5\ 4)$. The increasing subsequences $2\ 3\ 5$ and $2\ 3\ 4$
are maximal, therefore either $\mathrm{LIS}(\pi) = \{2, 3, 5\}$ or $\mathrm{LIS}(\pi) = \{2, 3, 4\}$.

3   Approximation Ratio Results

Benoît-Gagné and Hamel [2] showed that it is always possible to decrease by one unit the number of plateaux of a permutation by applying a transposition. Given that I_n is the only permutation having zero plateaux, an algorithm that decreases by one unit the number of plateaux of a permutation at each iteration sorts any permutation π with $p(\pi)$ transpositions. Although Benoît-Gagné and Hamel [2] stated that this algorithm, called *EasySorting* by them and hereafter also referred to as Algorithm 1, is a 3-approximation, we think that they did not provide a complete proof for such claim. That is, they proved that Algorithm 1 has the approximation factor given by Lemma 3, but this is not sufficient for proving an approximation factor bound of 3.

Lemma 3. *[2] An approximation factor of $d(\pi)$, for permutation π, with Algorithm 1 is*

$$c\frac{p(\pi)}{b(\pi)}, \ c = \frac{3\lfloor \frac{b(\pi)}{3} \rfloor + b(\pi) \mod 3}{\lceil \frac{b(\pi)}{3} \rceil}.$$

Since they showed that $c \leq 3$, it only lacks to prove that $p(\pi) \leq b(\pi)$. The proof is given by Lemma 4.

Lemma 4. *Given a permutation $\pi \in S_n$, we have that $p(\pi) < b(\pi)$.*

Proof. Let $\pi \in S_n$, $\pi \neq I_n$, s_1 be the first strip of π, and s_m be the last strip of π, $m \leq n$. If $\pi_1 = 1$, then $lc(\pi_i) = rc(\pi_i) = 0$ for any element $\pi_i \in s_1$. Thus, the elements belonging to s_1 do not affect $p(\pi)$. The same can be observed for the elements belonging to s_m when $\pi_n = n$. It means that if we reduce π to σ, then $p(\pi) = p(\sigma)$. Since it is not hard to see that $b(\pi) = b(\sigma)$, we can restrict our analysis to irreducible permutations.

Let $\gamma \in S_n^*$, and let the series of consecutive elements $\gamma_i \ \gamma_{i+1} \ \dots \ \gamma_j$ be a strip of γ. We have that $lc(\gamma_{k+1}) = lc(\gamma_k)$ and $rc(\gamma_{k+1}) = rc(\gamma_k)$, $i \leq k < j$. It means that, with respect to $lc(\gamma)$ and $rc(\gamma)$, the elements belonging to a strip of γ either have zero value or are contained in the same plateau. Therefore, $s(\gamma) \geq p(lc(\gamma))$ and $s(\gamma) \geq p(rc(\gamma))$, thus $s(\gamma) \geq p(\gamma)$. Since, by Lemma 2, $b(\gamma) > s(\gamma)$, the claim follows. $\qquad\square$

By using a structure based on breakpoints, Walter, Dias, and Meidanis [10] were able to develop an approximation algorithm that removes at least 4 breakpoints with 3 transpositions. Thus, if the algorithm sorts a permutation π with $a(\pi)$ transpositions, we have $a(\pi) \leq \frac{3}{4}b(\pi)$. Since a transposition can eliminate at most 3 breakpoints, we have $d(\pi) \geq \frac{b(\pi)}{3}$. Then, $\frac{a(\pi)}{d(\pi)} = \frac{9}{4}$ in the worst case, and the algorithm is therefore a 2.25-approximation. This algorithm will also be referred to as Algorithm 2 hereafter.

Guyer, Heath, and Vergara [9] developed a greedy algorithm based on the longest increasing subsequence of a permutation $\pi \in S_n$. At each iteration, the algorithm selects, from the $\binom{n}{3}$ possible transpositions, the transposition $\rho(i, j, k)$ such that $|\mathrm{LIS}(\rho(i, j, k) \cdot \pi)|$ is maximum. We say that a transposition satisfying this greedy choice is a greedy transposition. Since there may exist more than one greedy transposition, the performance of this algorithm may vary depending on the rule used to choose among greedy transpositions. Guyer, Heath, and Vergara [9] neither pointed out any specific rule nor presented an approximation guarantee, therefore we decided to define a rule that could lead one to determine an approximation guarantee, and then audit the algorithm yielded by this rule.

We say that a transposition $\rho(i, j, k)$ does not cut a strip of a permutation π if the pairs of adjacent elements (π_{i-1}, π_i), (π_{j-1}, π_j) and (π_{k-1}, π_k) are breakpoints. The rule we defined is that only greedy transpositions which do not cut strips of permutations must be applied. Algorithm 3 is the algorithm yielded by this rule. Lemma 5 shows that there is always a transposition satisfying such rule, while Theorem 1 proves that Algorithm 3 is a 3-approximation.

Data: Permutation $\pi \in S_n$
Result: Sequence of transpositions that sorts π

1   $d \leftarrow 0$;
2   **while** $\pi \neq I_n$ **do**
3      $d \leftarrow d + 1$;
4      $\rho_d \leftarrow$ the transposition such that $|\mathrm{LIS}(\rho_d \cdot \pi)|$ is maximum and that does not cut a strip of π. If there is more than one such transposition, apply the first one found;
5      $\pi \leftarrow \rho_d \cdot \pi$;
6   **end**
7   **return** $\rho_1, \rho_2, \ldots, \rho_d$

Algorithm 3. Constrained version of Guyer, Heath, and Vergara's algorithm [9]

Lemma 5. *Given a permutation π, there exists a greedy transposition which does not cut any of its strips.*

Proof. Let $\rho(i, j, k)$ be a greedy transposition, and let π' be the permutation such that $\pi' = \rho(i, j, k) \cdot \pi$. If $\rho(i, j, k)$ does not cut a strip of π, then we are done. Otherwise, we have to basically consider three possibilities:

(a) (π_{i-1}, π_i) is not a breakpoint.
   In this case let i' be the greatest integer such that $i' < i$ and $(\pi_{i'-1}, \pi_{i'})$ is a breakpoint, and let π'' be the permutation such that $\pi'' = \rho(i', j, k) \cdot \pi$. Then, we have four subcases to analyze:
   (i) $\pi_i \in \mathrm{LIS}(\pi')$ and $\{\pi_{i'}, \pi_{i'+1}, \ldots, \pi_{i-1}\} \in \mathrm{LIS}(\pi')$. In this subcase we have that $\{\pi_{i'}, \pi_{i'+1}, \ldots, \pi_{i-1}\} \in \mathrm{LIS}(\pi'')$, therefore $\rho(i', j, k)$ is also a greedy transposition once $|\mathrm{LIS}(\pi'')| = |\mathrm{LIS}(\pi')|$.
   (ii) $\pi_i \in \mathrm{LIS}(\pi')$ and $\{\pi_{i'}, \pi_{i'+1}, \ldots, \pi_{i-1}\} \notin \mathrm{LIS}(\pi')$. In this subcase we have that $\{\pi_{i'}, \pi_{i'+1}, \ldots, \pi_{i-1}\} \in \mathrm{LIS}(\pi'')$, therefore $|\mathrm{LIS}(\pi'')| > |\mathrm{LIS}(\pi')|$ and this contradicts our hypothesis that $\rho(i, j, k)$ is a greedy transposition.

(iii) $\pi_i \notin \mathrm{LIS}(\pi')$ and $\{\pi_{i'}, \pi_{i'+1}, \ldots, \pi_{i-1}\} \in \mathrm{LIS}(\pi')$. In this subcase we have that $|\mathrm{LIS}(\rho(i+1, j, k) \cdot \pi)| > |\mathrm{LIS}(\pi')|$ and this contradicts our hypothesis that $\rho(i, j, k)$ is a greedy transposition.

(iv) $\pi_i \notin \mathrm{LIS}(\pi')$ and $\{\pi_{i'}, \pi_{i'+1}, \ldots, \pi_{i-1}\} \notin \mathrm{LIS}(\pi')$. In this subcase we have that $\{\pi_{i'}, \pi_{i'+1}, \ldots, \pi_{i-1}\} \notin \mathrm{LIS}(\pi'')$, therefore $\rho(i', j, k)$ is also a greedy transposition once $|\mathrm{LIS}(\pi'')| = |\mathrm{LIS}(\pi')|$.

(b) (π_{j-1}, π_j) is not a breakpoint.

In this case let j' be the least integer such that $j < j'$ and $(\pi_{j'-1}, \pi_{j'})$ is a breakpoint, and let j'' be the greatest integer such that $j'' < j$ and $(\pi_{j''-1}, \pi_{j''})$ is a breakpoint. It may be the case that either $j' = k$ or $j'' = i$, but it is impossible that $j' = k$ and $j'' = i$, otherwise $\rho(i, j, k)$ would only move elements belonging to the same strip, therefore $|\mathrm{LIS}(\pi')| \leq |\mathrm{LIS}(\pi)|$ and $\rho(i, j, k)$ could not be a greedy transposition. Also note that a situation where $\pi_{j-1} \in \mathrm{LIS}(\pi')$ and $\pi_j \in \mathrm{LIS}(\pi')$ is not possible given the definition of an increasing subsequence. Then, if we assume that $j' \neq k$ and let π'' be the permutation such that $\pi'' = \rho(i, j', k) \cdot \pi$, we have three subcases to analyze:

(i) $\pi_{j-1} \in \mathrm{LIS}(\pi')$ and $\{\pi_j, \pi_{j+1}, \ldots, \pi_{j'-1}\} \notin \mathrm{LIS}(\pi')$. In this subcase we have that $\{\pi_j, \pi_{j+1}, \ldots, \pi_{j'-1}\} \in \mathrm{LIS}(\pi'')$, therefore $|\mathrm{LIS}(\pi'')| > |\mathrm{LIS}(\pi')|$ and this contradicts our hypothesis that $\rho(i, j, k)$ is a greedy transposition.

(ii) $\pi_{j-1} \notin \mathrm{LIS}(\pi')$ and $\{\pi_j, \pi_{j+1}, \ldots, \pi_{j'-1}\} \in \mathrm{LIS}(\pi')$. In this subcase we have that $|\mathrm{LIS}(\rho(i, j-1, k) \cdot \pi)| > |\mathrm{LIS}(\pi')|$ and this contradicts our hypothesis that $\rho(i, j, k)$ is a greedy transposition.

(iii) $\pi_{j-1} \notin \mathrm{LIS}(\pi')$ and $\{\pi_j, \pi_{j+1}, \ldots, \pi_{j'-1}\} \notin \mathrm{LIS}(\pi')$. In this subcase we have that $\{\pi_j, \pi_{j+1}, \ldots, \pi_{j'-1}\} \notin \mathrm{LIS}(\pi'')$, therefore $\rho(i, j', k)$ is also a greedy transposition once $|\mathrm{LIS}(\pi'')| = |\mathrm{LIS}(\pi')|$.

On the other hand, if we assume that $j'' \neq i$ and let π'' be the permutation such that $\pi'' = \rho(i, j'', k) \cdot \pi$, we also have three subcases to analyze:

(i) $\pi_j \in \mathrm{LIS}(\pi')$ and $\{\pi_{j''}, \pi_{j''+1}, \ldots, \pi_{j-1}\} \notin \mathrm{LIS}(\pi')$. In this subcase we have that $\{\pi_{j''}, \pi_{j''+1}, \ldots, \pi_{j-1}\} \in \mathrm{LIS}(\pi'')$, therefore $|\mathrm{LIS}(\pi'')| > |\mathrm{LIS}(\pi')|$ and this contradicts our hypothesis that $\rho(i, j, k)$ is a greedy transposition.

(ii) $\pi_j \notin \mathrm{LIS}(\pi')$ and $\{\pi_{j''}, \pi_{j''+1}, \ldots, \pi_{j-1}\} \in \mathrm{LIS}(\pi')$. In this subcase we have that $|\mathrm{LIS}(\rho(i, j+1, k) \cdot \pi)| > |\mathrm{LIS}(\pi')|$ and this contradicts our hypothesis that $\rho(i, j, k)$ is a greedy transposition.

(iii) $\pi_j \notin \mathrm{LIS}(\pi')$ and $\{\pi_{j''}, \pi_{j''+1}, \ldots, \pi_{j-1}\} \notin \mathrm{LIS}(\pi')$. In this subcase we have that $\{\pi_{j''}, \pi_{j''+1}, \ldots, \pi_{j-1}\} \notin \mathrm{LIS}(\pi'')$, therefore $\rho(i, j'', k)$ is also a greedy transposition once $|\mathrm{LIS}(\pi'')| = |\mathrm{LIS}(\pi')|$.

(c) (π_{k-1}, π_k) is not a breakpoint.

In this case let k' be the least integer such that $k < k'$ and $(\pi_{k'-1}, \pi_{k'})$ is a breakpoint, and let π'' be the permutation such that $\pi'' = \rho(i, j, k') \cdot \pi$. Then, we have four subcases to analyze:

(i) $\pi_{k-1} \in \mathrm{LIS}(\pi')$ and $\{\pi_k, \pi_{k+1}, \ldots, \pi_{k'-1}\} \in \mathrm{LIS}(\pi')$. In this subcase we have that $\{\pi_k, \pi_{k+1}, \ldots, \pi_{k'-1}\} \in \mathrm{LIS}(\pi'')$, therefore $\rho(i, j, k')$ is also a greedy transposition once $|\mathrm{LIS}(\pi'')| = |\mathrm{LIS}(\pi')|$.

(ii) $\pi_{k-1} \in \text{LIS}(\pi')$ and $\{\pi_k, \pi_{k+1}, \ldots, \pi_{k'-1}\} \notin \text{LIS}(\pi')$. In this subcase we have that $\{\pi_k, \pi_{k+1}, \ldots, \pi_{k'-1}\} \in \text{LIS}(\pi'')$, therefore $|\text{LIS}(\pi'')| > |\text{LIS}(\pi')|$ and this contradicts our hypothesis that $\rho(i, j, k)$ is a greedy transposition.

(iii) $\pi_{k-1} \notin \text{LIS}(\pi')$ and $\{\pi_k, \pi_{k+1}, \ldots, \pi_{k'-1}\} \in \text{LIS}(\pi')$. In this subcase we have that $|\text{LIS}(\rho(i, j, k-1) \cdot \pi)| > |\text{LIS}(\pi')|$ and this contradicts our hypothesis that $\rho(i, j, k)$ is a greedy transposition.

(iv) $\pi_{k-1} \notin \text{LIS}(\pi')$ and $\{\pi_k, \pi_{k+1}, \ldots, \pi_{k'-1}\} \notin \text{LIS}(\pi')$. In this subcase we have that $\{\pi_k, \pi_{k+1}, \ldots, \pi_{k'-1}\} \notin \text{LIS}(\pi'')$, therefore $\rho(i, j, k')$ is also a greedy transposition once $|\text{LIS}(\pi'')| = |\text{LIS}(\pi')|$.

Although more than one possibility can occur at the same time, they are independent from each other, in such a way that in all possible cases, if transposition $\rho(i, j, k)$ cuts a strip of π, then it is possible to derive a greedy transposition which does not, thus the claim follows. □

Theorem 1. *Algorithm 3 is a 3-approximation.*

Proof. For determining an upper bound to the number of transpositions applied by Algorithm 3, we define a simple algorithm, called *StripSum*, which receives as input a permutation $\pi \in S_n$, and proceeds as follows. Firstly, it sorts all the strips of π with respect to their sizes, obtaining a list of strips $s^0, s^1, \ldots, s^{s(\pi)-1}$ such that $|s^i| \geq |s^{i+1}|$ for all i, $0 \leq i < s(\pi) - 1$. Secondly, it initializes a variable named SUM to $|s^0|$. Finally, starting from s^1, it iterates over the list of strips such that, at iteration i, the algorithm increases the value of SUM by $|s^i|$. Let SUM_i be the value of the variable SUM at iteration i, with $\text{SUM}_0 = |s^0|$. Clearly, $\text{SUM}_i = \text{SUM}_{i-1} + |s^i|$ for all i, $1 \leq i \leq s(\pi) - 1$. Besides, when the algorithm stops, $\text{SUM} = \text{SUM}_{s(\pi)-1} = |s^0| + |s^1| + \cdots + |s^{s(\pi)}| = n$.

Now, assume that π was given as input to Algorithm 3, and let π^i be the permutation produced after i iterations, with $\pi^0 = \pi$. We can prove by induction that $|\text{LIS}(\pi^i)| \geq \text{SUM}_i$. Firstly, notice that $|\text{LIS}(\pi^0)| \geq \text{SUM}_0$ because, by definition, $|\text{LIS}(\pi^0)|$ must be equal or greater than the size of any strip of π^0. Then, assume $|\text{LIS}(\pi^k)| \geq \text{SUM}_k$ for all k, $0 \leq k \leq i$. Since Algorithm 3 never cuts a strip, all the strips of π^i are formed by strips of π^0. Let s' be the strip of greatest size among all strips of π^0 whose elements do not belong to a given $\text{LIS}(\pi^i)$. We have that $|\text{LIS}(\pi^{i+1})| \geq |\text{LIS}(\pi^i)| + |s'|$ because it is possible to apply a transposition to π^i and obtain a new permutation containing an increasing subsequence formed by the elements of s' and $\text{LIS}(\pi^i)$. If $|s'| \geq |s^{i+1}|$, then $|\text{LIS}(\pi^{i+1})| \geq |\text{LIS}(\pi^i)| + |s'| \geq \text{SUM}_i + |s^{i+1}| = \text{SUM}_{i+1}$. Otherwise, if $|s'| < |s^{i+1}|$, it means that the elements of all strips s^t, $0 \leq t \leq i+1$, belong to $\text{LIS}(\pi^i)$, therefore $|\text{LIS}(\pi^{i+1})| > |\text{LIS}(\pi^i)| \geq \text{SUM}_{i+1}$.

The inequality $|\text{LIS}(\pi^i)| \geq \text{SUM}_i$ implies that Algorithm 3 makes $|\text{LIS}(\pi)|$ converge to n applying no more transpositions than the number of iterations Algorithm StripSum performs. Thus, denoting by $A_3(\pi)$ the number of transpositions applied by Algorithm 3 for sorting π, we have $A_3(\pi) \leq s(\pi) - 1$.

Notice that if $\pi_1 = 1$, then the elements of the first strip will belong to $\mathrm{LIS}(\pi^i)$ for all i, $0 \leq i \leq A_3(\pi)$. This implies that Algorithm 3 will never apply a transposition which moves the elements of the first strip of π if $\pi_1 = 1$. The same can be observed for the elements of the last strip of π if $\pi_n = n$. For this reason, if we reduce permutation π to a permutation σ, it is not hard to see that $A_3(\pi) = A_3(\sigma)$. Thus, we can restrict our analysis to irreducible permutations.

Let $\gamma \in S_n^*$. We have that $A_3(\gamma) \leq s(\gamma) - 1$. Since, by Lemma 2, $s(\gamma) = b(\gamma)$ $- 1$, we conclude that $A_3(\gamma) \leq b(\gamma) - 2$. It means that $A_3(\gamma) \leq 3d(\gamma)$ once $d(\gamma)$ $\geq \frac{b(\gamma)}{3}$ (Lemma 1), and the theorem has been proved.   □

Since there are $O(n^3)$ possible transpositions, it takes $O(n\log n)$ time to determine a longest increasing subsequence of a permutation, it takes $O(1)$ time to determine whether a transposition cuts a strip of a permutation, and while loop executes $O(n)$ times, we conclude that Algorithm 3 runs in $O(n^5\log n)$ time.

4   Experimental Results and Discussion

The approximation algorithms presented in Sect. 3 were implemented and tested by their authors for verifying their performance in practice. One kind of test was to compare the distance computed by the algorithm with $d(\pi)$ for every $\pi \in S_n$ in order to obtain the approximation factor of the respective approximation algorithm for small permutations. More specifically, Walter, Dias, and Meidanis [10] ran this test for $1 \leq n \leq 11$, Benoît-Gagné and Hamel [2] ran it for $1 \leq n \leq 9$, and Guyer, Heath, and Vergara [9] ran it just for $n = 6$.

We ran this kind of test for $1 \leq n \leq 13$ for all algorithms using GRAAu [7], and the results are presented in tables 1, 2, and 3, where n is the size of the permutations, *Diameter* is the greatest distance outputted by the algorithm, *Avg. Distance* is the average of the distances outputted by the algorithm, *Avg. Ratio* is the average of the ratios between the distance outputted by the algorithm and the transposition distance, *Max. Ratio* is the greatest ratio among all the ratios between the distance outputted by the algorithm and the transposition distance, and *Equals* is the percentage of distances outputted by the algorithm that is equal to the transposition distance.

Regarding the 3-approximation algorithm developed by Benoît-Gagné and Hamel [2], if we just consider the approximation factors obtained for $n \in \{7, 10, 13\}$, we can observe that they seem to follow the progression $\frac{6}{3}, \frac{9}{4}, \frac{12}{5}, \ldots, \frac{3k}{k+1}$. We ran further experiments to verify the strength of this assumption, and we found permutations π_m of size $3m + 1$, $m \in \{5, 6, 7\}$, for which $\frac{p(\pi_m)}{d(\pi_m)} = \frac{3m}{m+1}$ (these permutations are presented in Table 4). This result strongly indicates that the approximation factor of this algorithm is tight, contradicting the hypothesis raised by Benoît-Gagné and Hamel [2] that its approximation factor "tends to a number significantly smaller than 3".

We note that the results obtained by Walter, Dias, and Meidanis [10] for their 2.25-approximation algorithm are incorrect. For instance, the maximum value of $\frac{a(\pi)}{d(\pi)}$ they observed for $n = 11$ was $\frac{10}{5}$. But this result cannot be right because

$a(\pi) \leq 9$ for every $\pi \in S_{11}$. Given a permutation $\pi \in S_n$, it is easy to see that $0 \leq b(\pi) \leq n+1$. As discussed in Sect. 2, $a(\pi) \leq \frac{3}{4}b(\pi)$, therefore $a(\pi) \leq \frac{3n+3}{4}$, and the claim follows. Since the values we obtained for $a(\pi)$ are consistent with the breakpoint upper bound, we conclude that the approximation factors presented in Table 2 are correct.

The approximation factors observed for the 2.25-approximation algorithm seem to increase in a progression that converges to 2, that is, $\frac{2}{2}, \frac{4}{3}, \frac{6}{4}, \frac{8}{5}, \dots,$ $\frac{2k}{k+1}$. This may indicate that a deeper analysis of this algorithm could lead one to prove that it is in fact a 2-approximation.

Figure 1 illustrates that, of all the algorithms analyzed in this paper, Walter, Dias, and Meidanis' algorithm [10] (Algorithm 2) has not only the best theoretical performance, but also the best practical performance.

Table 1. Results obtained from the audit of the implementation of Benoît-Gagné and Hamel's algorithm [2]

n	Diameter	Avg. Distance	Avg. Ratio	Max. Ratio	Equals
1	0	0.00	1.00	1.00	100.00%
2	1	0.50	1.00	1.00	100.00%
3	2	1.00	1.00	1.00	100.00%
4	3	1.54	1.00	1.00	100.00%
5	4	2.13	1.02	1.50	95.00%
6	5	2.75	1.06	1.67	85.00%
7	6	3.42	1.10	2.00	71.77%
8	7	4.13	1.14	2.00	56.41%
9	8	4.87	1.18	2.00	41.62%
10	9	5.63	1.22	2.25	28.80%
11	10	6.42	1.25	2.25	18.74%
12	11	7.22	1.29	2.25	11.57%
13	12	8.05	1.32	2.40	6.77%

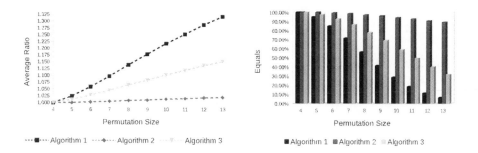

Fig. 1. Performance comparison of Benoît-Gagné and Hamel's algorithm [2] (Algorithm 1), Walter, Dias, and Meidanis' algorithm [10] (Algorithm 2), and the constrained version of Guyer, Heath, and Vergara's algorithm [9] (Algorithm 3) based on the results provided by GRAAu [7]

Table 2. Results obtained from the audit of the implementation of Walter, Dias, and Meidanis' algorithm [10]

n	Diameter	Avg. Distance	Avg. Ratio	Max. Ratio	Equals
1	0	0.00	1.00	1.00	100.00%
2	1	0.50	1.00	1.00	100.00%
3	2	1.00	1.00	1.00	100.00%
4	3	1.54	1.00	1.00	100.00%
5	3	2.08	1.00	1.00	100.00%
6	4	2.61	1.00	1.33	99.17%
7	5	3.14	1.00	1.33	98.57%
8	6	3.66	1.01	1.50	97.12%
9	6	4.19	1.01	1.50	96.06%
10	7	4.70	1.01	1.50	94.15%
11	8	5.22	1.01	1.60	92.84%
12	9	5.73	1.02	1.60	90.68%
13	9	6.24	1.02	1.60	89.30%

Table 3. Results obtained from the audit of the implementation of Algorithm 3, which is a constrained version of Guyer, Heath, and Vergara's algorithm [9]

n	Diameter	Avg. Distance	Avg. Ratio	Max. Ratio	Equals
1	0	0.00	1.00	1.00	100.00%
2	1	0.50	1.00	1.00	100.00%
3	2	1.00	1.00	1.00	100.00%
4	3	1.54	1.00	1.00	100.00%
5	4	2.10	1.01	1.50	97.50%
6	5	2.67	1.03	1.50	92.78%
7	6	3.26	1.05	1.67	86.45%
8	7	3.86	1.06	1.67	77.93%
9	8	4.48	1.08	2.00	69.06%
10	9	5.10	1.10	2.00	58.94%
11	10	5.73	1.12	2.00	49.61%
12	11	6.38	1.14	2.00	40.23%
13	12	7.03	1.15	2.25	32.18%

Table 4. Permutations π_m of size $3m+1$, $m \in \{5, 6, 7\}$, for which $\frac{p(\pi_m)}{d(\pi_m)} = \frac{3m}{m+1}$. Note that $d(\pi_m) \geq \frac{b(\pi_m)}{3} \geq m + 1$

Permutation	Transposition Sorting Sequence
$\pi_5 = (16\ 9\ 4\ 11\ 6\ 15\ 8\ 2\ 12\ 7\ 5\ 3\ 14$ $13\ 10\ 1)$	$\rho(5, 9, 12)$, $\rho(1, 7, 10)$, $\rho(3, 9, 14)$, $\rho(5, 10, 17)$, $\rho(6, 10, 14)$, $\rho(1, 7, 11)$
$\pi_6 = (19\ 11\ 4\ 18\ 6\ 14\ 8\ 13\ 10\ 2\ 15$ $5\ 9\ 7\ 3\ 17\ 16\ 12\ 1)$	$\rho(4, 12, 17)$, $\rho(7, 12, 16)$, $\rho(6, 9, 15)$, $\rho(1, 5, 11)$, $\rho(3, 8, 20)$, $\rho(4, 13, 17)$, $\rho(1, 5, 13)$
$\pi_7 = (22\ 13\ 4\ 21\ 6\ 17\ 8\ 16\ 10\ 15\ 12$ $2\ 18\ 5\ 11\ 9\ 7\ 3\ 20\ 19\ 14\ 1)$	$\rho(4, 14, 20)$, $\rho(8, 13, 19)$, $\rho(7, 10, 18)$, $\rho(6, 9, 17)$, $\rho(1, 5, 11)$, $\rho(3, 8, 23)$, $\rho(4, 16, 20)$, $\rho(1, 5, 15)$

5   Conclusions

We conducted an experimental investigation on the performance of the 3-approximation algorithm developed by Benoît-Gagné and Hamel [2], of the 2.25-approximation algorithm presented by Walter, Dias, and Meidanis [10], and of a constrained version of Guyer, Heath, and Vergara's heuristic [9] for sorting by transpositions. We verified that the approximation factor of the first two seem to increase in a progression that converges to 3 and to 2 respectively, what suggests that the 2.25-approximation algorithm may be a 2-approximation, and that the approximation ratio of Benoît-Gagné and Hamel's algorithm [2] can be as close to 3 as we want. Besides, we proved an approximation bound of 3 for the constrained version of Guyer, Heath, and Vergara's heuristic [9], and the experimental results we obtained indicate that this bound cannot be lower than 2.25.

Although we did not present any formal proof of the tightness of these algorithms at this time, our results suggest that the approaches considered by them are not promising alternatives in the design of approximation algorithms with low approximation ratios. They do not seem to be good alternatives even when it comes to practical performance, excepting Walter, Dias, and Meidanis' algorithm [10], which presents relatively good results compared to theoretically better algorithms, as shown by Dias and Dias [4].

We hope that we can use the approaches we [8] used to prove the tightness of the 2-approximation algorithms for Sorting by Prefix Reversals and Sorting by Prefix Transpositions for proving the tightness of the approximation algorithms discussed in this work.

Acknowledgments. We thank the support of National Council for Scientific and Technological Development (process 473867/2010-9). We also thank the anonymous reviewers for their constructive comments.

References

1. Bafna, V., Pevzner, P.A.: Sorting by transpositions. SIAM Journal on Discrete Mathematics 11(2), 224–240 (1998)
2. Benoît-Gagné, M., Hamel, S.: A New and Faster Method of Sorting by Transpositions. In: Ma, B., Zhang, K. (eds.) CPM 2007. LNCS, vol. 4580, pp. 131–141. Springer, Heidelberg (2007)
3. Bulteau, L., Fertin, G., Rusu, I.: Sorting by Transpositions Is Difficult. In: Aceto, L., Henzinger, M., Sgall, J. (eds.) ICALP 2011, Part I. LNCS, vol. 6755, pp. 654–665. Springer, Heidelberg (2011)
4. Dias, U., Dias, Z.: Extending Bafna-Pevzner algorithm. In: Proceedings of the International Symposium on Biocomputing (ISB 2010), pp. 1–8. ACM, New York (2010)
5. Elias, I., Hartman, T.: A 1.375-approximation algorithm for sorting by transpositions. IEEE/ACM Transactions on Computational Biology and Bioinformatics (TCBB) 3(4), 369–379 (2006)

6. Fertin, G., Labarre, A., Rusu, I., Tannier, E., Vialette, S.: Combinatorics of Genome Rearrangements. The MIT Press (2009)
7. Galvão, G.R., Dias, Z.: GRAAu: Genome Rearrangement Algorithm Auditor. In: Proceedings of the 4th International Conference on Bioinformatics and Computational Biology (BICoB 2012), Las Vegas, NV, USA, pp. 97–101 (2012)
8. Galvão, G.R., Dias, Z.: On the performance of sorting permutations by prefix operations. In: Proceedings of the 4th International Conference on Bioinformatics and Computational Biology (BICoB 2012), Las Vegas, NV, USA, pp. 102–107 (2012)
9. Guyer, S.A., Heath, L.S., Vergara, J.P.: Subsequence and run heuristics for sorting by transpositions. Technical Report TR-97-20, Computer Science, Virginia Polytechnic Institute and State University (1997)
10. Walter, M.E.M.T., Dias, Z., Meidanis, J.: A new approach for approximating the transposition distance. In: Proceedings of the Seventh International Symposium on String Processing Information Retrieval (SPIRE 2000), pp. 199–208. IEEE Computer Society, Washington, DC (2000)

A Comparison of Three Heuristic Methods
for Solving the Parsing Problem
for Tandem Repeats

A.A. Matroud[1,3], C.P. Tuffley[1], D. Bryant[2,3], and M.D. Hendy[2]

[1] Institute of Fundamental Sciences, Massey University, Private Bag 11222,
Palmerston North, New Zealand
[2] Department of Mathematics and Statistics, University of Otago, Dunedin,
New Zealand
[3] Allan Wilson Centre for Molecular Ecology and Evolution

Abstract. In many applications of tandem repeats the outcome depends critically on the choice of boundaries (beginning and end) of the repeated motif: for example, different choices of pattern boundaries can lead to different duplication history trees. However, the best choice of boundaries or *parsing* of the tandem repeat is often ambiguous, as the flanking regions before and after the tandem repeat often contain partial approximate copies of the motif, making it difficult to determine where the tandem repeat (and hence the motif) begins and ends. We define the *parsing problem* for tandem repeats to be the problem of discriminating among the possible choices of parsing.

In this paper we propose and compare three heuristic methods for solving the parsing problem, under the assumption that the parsing is fixed throughout the duplication history of the tandem repeat. The three methods are PAIR, which minimises the number of pairs of common adjacent mutations which span a boundary; VAR, which minimises the total number of variants of the motif; and MST, which minimises the length of the minimum spanning tree connecting the variants, where the weight of each edge is the Hamming distance of the pair of variants. We test the methods on simulated data over a range of motif lengths and relative rates of substitutions to duplications, and show that all three perform better than choosing the parsing arbitrarily. Of the three MST typically performs the best, followed by VAR then PAIR.

1 Introduction

Genomic DNA has long been known to contain *tandem repeats:* repetitive structures in which many approximate copies of a common segment (the *motif*) appear consecutively. The copies of the motif are usually polymorphic, which makes tandem repeats a useful tool for phylogenetics and for inter-population studies (Rivals [15]); in addition, highly polymorphic tandem repeats can be used to discriminate among individuals within a population, and have proved to be useful for DNA fingerprint techniques (Jeffreys et al. [11]). Because of this, many

M.C.P. de Souto and M.G. Kann (Eds.): BSB 2012, LNBI 7409, pp. 37–48, 2012.

algorithms have been developed to find tandem repeats; align tandem repeats; compare DNA sequences containing tandem repeats (also known as mapping tandem repeats; Behzadi and Steyaert [1]; Berard and Rivals [4]; Sammeth and Stoye [17]); and construct the duplication history tree (DHT) of a tandem repeat (Rivals [15]; Lajoie et al. [13]; Bertrand et al. [5]; Chauve et al. [6]).

In many important applications of tandem repeats the outcome depends critically on the choice of boundaries (beginning and end) of the repeated motif. We will refer to a choice of boundaries as a *parsing* of the tandem repeat. For a tandem repeat with motif of length ℓ there are ℓ possible parsings, and different choices of parsing can for example lead to different duplication history trees. However, the "true" parsing is often ambiguous, as the flanking regions (the ℓ nucleotides immediately preceding and following the tandem repeat) often contain partial approximate copies of the motif, making it difficult to decide where the tandem repeat (and consequently the motif) begins and ends. It is therefore highly desirable to find methods to discriminate among the possible parsings, and we will refer to the problem of doing so as the *parsing problem* for tandem repeats.

The parsing problem does not appear to have received a great deal of attention to date in the literature, and in many tandem repeat search tools the criteria for setting boundaries appear to be subjective or arbitrary (for example Crochemore [7]; Matroud et al. [14]; Hauth and Joseph [10]; Stoye and Gusfield [18]; Benson [2]; Sagot and Myers [16]). To the best of our knowledge the only reference on this problem to date is by Benson and Dong [3], who propose a method to solve the parsing problem based on a tandem repeat duplication model which allows dynamic boundaries (that is, duplications may occur on different boundaries throughout the duplication history). The purpose of this paper is to present three new criteria to select the parsing, under the assumption that the pattern boundaries are fixed throughout the duplication process (Fitch [8]).

One possible method for selecting the parsing is to choose the parsing that minimises the parsimony score of the resulting DHT. However, obtaining the maximum parsimony DHT can be computationally expensive (especially when the motif is long or there are many copies), and in some cases the maximum parsimony tree cannot be expressed as a duplication tree (Gascuel et al. [9]). In these circumstances it may be preferable to find more tractable measures for comparing parsings. The three criteria we propose are heuristic, and are intended as easily computed surrogates for the score of the maximum parsimony tree. They are each based on the observation that the histories of two nucleotide substitutions that occur at nearby sites less than ℓ bases apart at some stage in the evolution of the tandem repeat can be different depending on whether they occur in the same or adjacent copies of the motif.

The three methods are

1. PAIR, which minimises the number of pairs of common adjacent mutations which span a boundary,

2. VAR, which minimises the total number of variants of the motif, and

3. MST, which minimises the length of the minimum spanning tree connecting the variants, with each edge length being the Hamming distance of the pair of variants.

We test the methods on simulated data (for which the parsing used to generate the tandem repeat is known) over a range of motif lengths and relative rates of substitutions to duplications. We show that all three methods perform better than choosing the parsing arbitrarily, and that of the three MST typically performs the best, followed by VAR then PAIR.

2   Definitions and Background

An *exact tandem repeat* is a string comprising two or more contiguous exact copies of a substring $\mathbf{X}$, called the tandem repeat *motif*. We obtain an *approximate tandem repeat* by allowing approximate rather than exact copies of the template motif $\mathbf{X}$. We will refer to each copy of the motif as a *segment*. We define a *mode* motif to be a sequence of length ℓ where the i−th nucleotide is a most common nucleotide at the i−th site among the segments, for all $1 \leq i \leq \ell$. We define the set of *variants* of a tandem repeat with motif $\mathbf{X}$ to be the set of distinct segments of the repeat. Let θ denote a transformation $\theta \in \{\alpha, \beta, \gamma\}$, where, following the Kimura 3ST substitution model [12], the transformation types are

$$\alpha = \text{A} \leftrightarrow \text{G, C} \leftrightarrow \text{T}; \quad \beta = \text{A} \leftrightarrow \text{T, G} \leftrightarrow \text{C}; \quad \gamma = \text{A} \leftrightarrow \text{C, G} \leftrightarrow \text{T}.$$

The *substitution $i\theta$* in a segment S denotes θ applied to the nucleotide at the site i in S.

We define the *distance graph* of the variants to be the weighted graph with vertex set the set of variants, and an edge between each pair of variants with weight equal to their Hamming distance. Note that the distance graph of the variants is a weighted complete graph.

Let $\mathbf{X}[i] = x_i x_{i+1} \ldots x_\ell x_1 \ldots x_{i-1}$ be the ith cyclic permutation of the motif $\mathbf{X}$, where ℓ is the length of $\mathbf{X}$. We define the *parsing problem* for tandem repeats to be the problem of determining which of the ℓ possible cyclic permutations of $\mathbf{X}$ is the best estimation of the ancestral segment parsing.

3   A Duplication Model for Tandem Repeats

Tandem repeats can be modeled as a consequence of single nucleotide substitutions, and duplications and deletions of one or more copies of the motif. A duplication event occurs when a segment of one or more motif copies is repeated adjacent to the original. The number of motifs copied is referred to as the size of the duplication. We assume that the boundaries of a duplication event fall on the boundaries of a repeat.

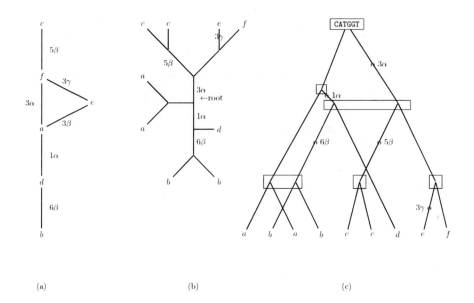

<table>
(a)   (b)   (c)
</table>

Fig. 1. In (a) we see a graph connecting the 6 variants in (2), with each edge u, v labelled by the substitution $i\theta$ that transforms u to v. (b) Arbitrarily breaking the cycle, and adding leaves for the multiple segments, we obtain a maximum parsimony tree of the 9 segments. When we place the root of the tree on the edge (arrowed) and order the tips as $ababccdef$ we obtain the duplication tree (c), descending from the modal motif a. There are 6 duplications, with the rectangles enclosing the segments being duplicated.

In our model, it is also assumed that duplication and substitution events occur at a fixed relative rate, and that the motif copies remain contiguous and oriented in the same direction in the genome. Under these assumptions the duplication history of a tandem repeat can be described by a DHT.

Example 1. We illustrate how a DHT may be inferred from a tandem repeat with a chosen parsing. Consider the following sequence which contains the tandem repeat

$$\text{TTATGT}\boxed{\text{CATGGT}}\text{TATGGA}\boxed{\text{CATGGT}}\text{TATGGA}\boxed{\text{CACGCT}}\text{CACGCT}\text{TATGGT}\boxed{\text{CAAGGT}}\boxed{\text{CACGGT}}\text{CAATAG},$$

$$(1)$$

which for the parsing displayed, is an approximate tandem repeat with modal motif CATGGT. There are six motif variants, in order as $ababccdef$ where

$$a = \text{CATGGT}, \ b = \text{TATGGA}, \ c = \text{CACGCT}, \ d = \text{TATGGT}, \ e = \text{CAAGGT}, \ f = \text{CACGGT}.$$

$$(2)$$

These variants may be represented by the graph in Figure 1(a), in which the edge between variants u and v is labelled by the substitution $i\theta$ that transforms u into v. Figure 1(c) shows a DHT in which each edge is labeled with zero or more substitutions of the form $i\theta$.

By removing one edge of the $a-e-f$ cycle (edge $a-e$ was chosen arbitrarily) in Figure 1(a) and adding leaves for each of the 9 segments, we obtain the maximum parsimony tree in Figure 1(b). In Figure 1(b) if we place the root of the tree on the edge arrowed, we get a DHT (this edge is the only edge where a root can be placed to get a duplication tree (Gascuel et al. [9])). Each duplication is identified by the segment to be duplicated enclosed in a rectangle. When the duplicated segment encloses more than one copy of the motif, the descendant motifs alternate as shown in Figure 1(c). The approximate tandem repeat is fully described by the duplication tree T with 5 duplications, the ancestral motif at the root (CATGGT), and the 5 substitutions on the edges of T.

4   The Importance of the Parsing Problem

In order to use a tandem repeat region in a phylogenetic study we need to infer the number and size of the edit operations that occurred in this region, transforming a single copy of the motif into the observed tandem repeat. However, the number and size of the inferred edit operations depend on the parsing we select. In the following example, we illustrate the implications of having two different parsing points on the inferred number and size of the edit operations.

Example 2. Consider the following sequence which contains an approximate tandem repeat with periodicity 4:

$$\text{AGACCACGAACGTACGAACGTATTA.} \tag{3}$$

There are 4 possible parsings. For each parsing, we define a *modal* motif to be a sequence where the i−th nucleotide is a most common nucleotide at the i−th site among the segments, for all $1 \leq i \leq \ell$. If we set the first boundary point such that the first repeat copy is ACGA we obtain

$$\text{AGACC}\boxed{\text{ACGA}}\boxed{\text{ACGT}}\boxed{\text{ACGA}}\boxed{\text{ACGT}}\text{ATTA,} \tag{4}$$

with mode motif ACGA (note that in this case ACGT is another modal motif). If we shift the frame one nucleotide to the left we obtain a different parsing

$$\text{AGAC}\boxed{\text{CACG}}\boxed{\text{AACG}}\boxed{\text{TACG}}\boxed{\text{AACG}}\text{TATTA,} \tag{5}$$

with a unique modal motif AACG.

The DHT generating the tandem repeat depends on the parsing. The DHT for the parsing of (4) is shown in Figure 2(a), and the DHT for the parsing of (5) is shown in Figure 2(b). The two trees involve different sets of edit operations; the first requires fewer, and so is to be preferred on parsimony grounds. Note that the parsing with modal motif CGAA that results from a frame shift one nucleotide to the right in (4) gives the same DHT as Figure 2(a).

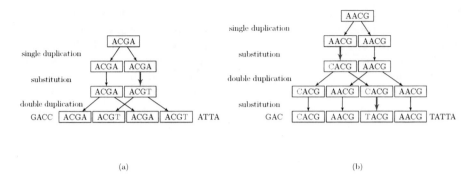

Fig. 2. The DHTs inferred from the two parsings of Example 2. The parsing (a) has a DHT with two duplications and a single substitution, in parsing (b) the DHT has two duplications and two substitutions. In both cases we see that the number of events is minimal for that parsing. By the parsimony principle, we prefer parsing (a) over parsing (b) as its DHT requires fewer mutational events.

5   Heuristic Methods to Estimate Tandem Repeat Parsing

In Example 2 we were able to discriminate between the parsings of (4) and (5) on the basis of the parsimony scores of their duplication history trees. However, when considering large and long tandem repeats, obtaining the maximum parsimony duplication history tree of the motif copies can be computationally expensive, and in some cases the maximum parsimony tree cannot be expressed as a duplication tree (Gascuel et al. [9]). It may be preferable to avoid these constructions when comparing different parsings.

Below, we describe three heuristic approaches to discriminating between the different possible parsings. They are each based on the observation that the histories of two nucleotide substitutions that occur at nearby sites less than ℓ bases apart at some stage in the evolution of the tandem repeat can be different depending on whether they occur in the same segment or in adjacent segments.

Suppose a substitution $i\theta_1$ producing nucleotide X occurs at site i in one segment in the sequence. Recall that $i\theta$ denotes a substitution of the type $\theta \in \{\alpha, \beta, \gamma\}$ at site i. The variant containing X may be duplicated a number of times before a second substitution $j\theta_2$ producing nucleotide Y occurs at site j within ℓ bases of X at site $i_1 \equiv i \bmod \ell$, with $j = i_1 + k$, $0 < |k| < \ell$. Now suppose there are further duplications producing further copies of variants containing X and Y.

If Y is in the same segment as X, and there are no subsequent parallel substitutions producing Y at any other site $j_1 \equiv j \bmod \ell$, then we will observe variants with X, variants with X and Y together, but no variants with Y alone. Hence each Y at site j_2 will have a companion X at site $i_2 = j_2 - k$. Figure 3 illustrates this scenario.

However, if the substitution producing Y were in an adjacent segment, and there were subsequent duplications, then we can observe variants containing X

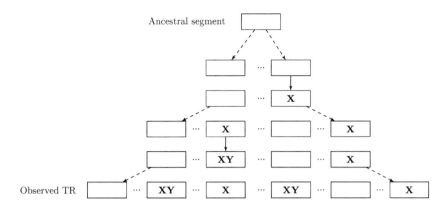

Fig. 3. A duplication tree. The dashed arrows represent series of duplication events of different sizes, and the solid arrows represent single nucleotide substitutions.

alone, and variants containing Y alone. In some of these cases there may be pairs of X and Y still k nucleotides apart, but there can also be copies of Y with no adjacent X. For the purpose of the arguments, we will assume that X is to the left of Y in the segments that contain both.

The three methods we have introduced to exploit this observation are listed below. We note the performance of these methods is dependent on the ratio m of substitutions to duplications. If m is small there will only be a small number of (or possibly no) pairs of close substitutions to indicate the likely parsing. In these cases, there may be multiple locations that are optimal on some of our scores, and some additional criteria (or perhaps random choice) would be required to identify a preferred parsing.

5.1   PAIR — The Adjacent Pairs Method

Here we consider all occurrences of a pair of substitutions X and Y which occur at least twice (to restrict attention to pairs which may have been duplicated) at sites less than ℓ bases apart, with the X's at some sites i mod ℓ, and the Y's at some sites j mod ℓ, and with each Y always adjacent to an X. We then note all the site gaps mod ℓ between each adjacent X and Y and record their frequency. For the method PAIR we select those site gaps mod ℓ which are counted in this way with lowest frequency as our preferred location for the boundary of the motifs. Provided there are sufficient substitutions so that multiple substitutions occur in some ancestral segment which is subsequently duplicated, then this should discriminate between site gaps. This discrimination should remain as the ratio m of substitutions to duplications grows, as the frequency of parallel substitutions should always be lower than unique substitutions.

To illustrate this method, consider the tandem repeat of Example 1 in Section 3. Indexing from site 1 in the first box, the substitutions from the modal motif are: $7\alpha, 12\beta, 19\alpha, 24\beta, 27\alpha, 29\beta, 33\alpha, 35\beta, 37\alpha, 45\beta, 51\alpha$. From this list we observe the only sets of pairs at most 5 bases apart which occur more than once

Table 1. The score under each of our three methods for each parsing of the tandem repeat in Example 1 in Section 3. See Sections 5.1 to 5.3 for details. Each method returns the first parsing CATGGT as the preferred parsing, since this minimises the score with respect to each method.

Method	Consensus pattern of parsing					
	CATGGT	TCATGG	GTCATG	GGTCAT	TGGTCA	ATGGTC
PAIR	0	2	2	4	4	2
VAR	6	7	8	7	7	7
MST	5	6	7	7	6	6

are $(1\alpha, 6\beta)$ at sites $(7, 12)$ and $(19, 24)$, and $(3\alpha, 5\beta)$ at sites $(27, 29)$ and $(33, 35)$. These pairs do not straddle the parsing boundary in (1), but either one or both pairs will straddle any other proposed parsing boundary, and contribute 2 (the frequency of the pair) to the score of the corresponding parsing. The resulting score for each parsing is shown in Table 1. The method selects the generating parsing CATGGT (1) as the preferred parsing.

5.2   VAR — The Number of Variants Method

For this method we consider each of the ℓ possible parsings, and for each parsing we count the number of variants. When the proposed parsing is correct, then the variants containing X and containing X and Y together will be counted, leading to 2 additional variants observed. However, if we propose a parsing which separates these, then we may find variants containing X alone, variants containing Y alone, and variants containing both, with Y to the left of X. Hence the VAR method selects the parsing or parsings which minimise the number of distinct variants.

As the ratio m of substitutions to duplications grows, so too will the number of distinct variants, and for m large, they may be almost all distinct, irrespective of the location of the proposed parsing. Hence for larger values of m, the VAR method may lose its discriminatory power.

To illustrate the method we again consider the tandem repeat in Example 1 in Section 3. For the parsing shown there are six variants, as listed in (2), so the score for this parsing is 6. The scores for the other six parsings are given in Table 1. The method selects the first parsing CATGGT as the preferred parsing, as this minimises the score.

5.3   MST — The Minimum Spanning Tree Method

When m is small, so the number of variants is small, then the Maximum Parsimony (MP) tree connecting the variants is likely to be 1−connected, and the length of the MP tree will be the number of variants minus one. However as m grows, the MP tree may require Steiner points (representing ancestral variants that are no longer present in the extant set of variants), and the length of the MP tree may be

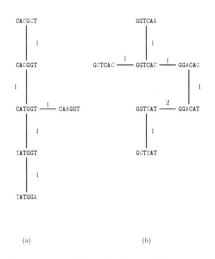

(a)   (b)

Fig. 4. Minimum spanning trees of the variants distance graphs for two of the six parsings of the tandem repeat in Example 1, Section 3. (a) A minimum spanning tree of length 5 for the parsing `CATGGT`. (b) A minimum spanning tree of length 7 for the parsing `GGTCAT`.

more discriminating in determining the parsing. However, as the determination of the MP tree can be NP-hard, we can use the length of the minimum spanning tree, as a quick measure of the relative relatedness of the variants. To avoid the issue of connectedness and the requirement of Steiner points, we take the minimum spanning tree of the distance graph of the variants. Recall that this is the graph with vertex set the set of variants, and an edge between two variants a and b with weight equal to their Hamming distance.

We propose the MST, the length of the minimum spanning tree of the variants as our third measure. We expect MST to agree with VAR, but to be more accurate for larger values of m.

We illustrate this method in Figure 4, which shows minimum spanning trees of the variants distance graphs for two of the six parsings of the tandem repeat in Example 1, Section 3. The first parsing `CATGGT` has a minimum spanning tree of length 5, while the fourth parsing `GGTCAT` has a minimum spanning tree of length 7. Note that in this latter case the minimum spanning tree length is greater than VAR -1, illustrating the potential for MST to be more discriminating than VAR. The scores for all six parsings appear in Table 1, and the method again selects the first parsing `CATGGT` as the preferred parsing.

6   Results and Discussion

We generated 90000 synthetic DNA sequences to compare the accuracy of the three proposed parsing methods. Each simulated sequence contained an approximate tandem repeat of around 100 copies of an ancestral motif of ℓbp ($\ell = 10, 50, 100$). These were generated by a stochastic evolutionary process of motif duplication (where the frequency of duplicating a segment of κ motif copies was proportional to $\frac{1}{\kappa}$), with nucleotide substitutions accumulating at a frequency of m substitutions per duplication. We applied each of the three proposed parsing methods to each simulated sequence and recorded whether the predicted parsing agreed with the parsing used to generate the tandem repeat.

We generated 100 samples for each value of $m = 0.1, 0.2, \ldots, 30.0$ and we report the percentage success for each method and value of m in Figure 5. For the purposes of this plot, "success" means that the set of minima reported by the method contains the true parsing, or one of the two parsings a step away from the true parsing. The average number of minima returned by each method is plotted in Figure 7 (plotted as a percentage of the number of possible parsings (the motif length ℓ)), which shows that each method typically returned only a small fraction of the possible parsings — often only one or two in the case of PAIR and MST.

We note that the PAIR method performed poorly (at about 35% accuracy) over the range of values of m, whereas VAR and MST showed above 90% accuracy for m in the range of about 0.3 to 5. Nevertheless all three methods performed better than setting the parsing arbitrarily, as the null method of randomly assigning a parsing would be expected to achieve accuracy of $3/\ell = 30\%, 6\%, 3\%$ for $\ell = 10, 50, 100$. We also note that the PAIR method has lower sensitivity to the motif length than the other two methods. For much of this range the MST method performs better than VAR. The VAR method does not perform better than MST in terms of accuracy (Figures 5 and 6), and it also produces solutions containing a larger number of optimal parsings (Figure 7).

Figure 6 shows the percentage failure. In this figure, we count the number of times the true parsing is not among those reported. The MST method shows less than 20% failure on average. With low values of m, there is insufficient variation to distinguish between the many possible parsings. Because of the low level of variation, we are likely to find many alternate parsing of similar score.

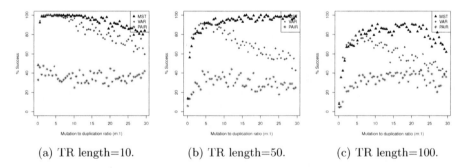

(a) TR length=10.   (b) TR length=50.   (c) TR length=100.

Fig. 5. Percentage success plotted against the relative mutation rate for each of the three methods. The y-axis represents the percentage of simulations for which the set of minima contains the true boundary point or the points that are one step away from the true boundary, plotted against each relative mutation rate $m = 0.1, 0.2, \ldots, 30$. The motif length is (a) $\ell = 10$, (b) $\ell = 50$ and (c) $\ell = 100$.

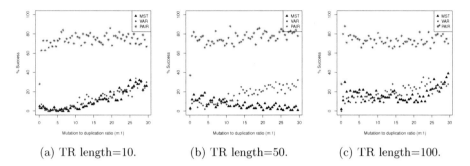

(a) TR length=10.   (b) TR length=50.   (c) TR length=100.

Fig. 6. Percentage failure plotted against the relative mutation rate for each of the three methods. The y-axis represents the percentage of simulations for which the true boundary is not among the reported minima, plotted against each relative mutation rate $m = 0.1, 0.2, \ldots, 30$. The motif length is (a) $\ell = 10$, (b) $\ell = 50$ and (c) $\ell = 100$.

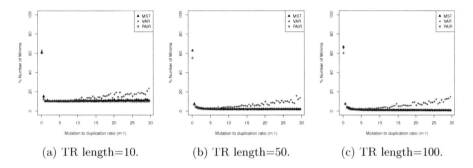

(a) TR length=10.   (b) TR length=50.   (c) TR length=100.

Fig. 7. The number of minima plotted against the relative mutation rate. The y-axis represents the average number of minima reported by each method, expressed as a percentage of the number of possible parsings (the motif length ℓ), plotted against each relative mutation rate $m = 0.1, 0.2, \ldots, 30$. The motif length is (a) $\ell = 10$, (b) $\ell = 50$ and (c) $\ell = 100$.

The PAIR method has the computational advantage that the scores for all ℓ possible parsings can be computed simultaneously, whereas the VAR and MST scores must be computed for each possible parsing in turn. Nevertheless, their observed accuracy indicates they are preferred to the PAIR method, when the motif length is not large. MST performed better on average than VAR, so our results suggest MST to be the preferred method of predicting parsing.

References

1. Behzadi, B., Steyaert, J.-M.: An Improved Algorithm for Generalized Comparison of Minisatellites. In: Baeza-Yates, R., Chávez, E., Crochemore, M. (eds.) CPM 2003. LNCS, vol. 2676, pp. 32–41. Springer, Heidelberg (2003)

2. Benson, G.: Tandem repeats finder: a program to analyze DNA sequences. Nucl. Acids Res. 27(2), 573–580 (1999)
3. Benson, G., Dong, L.: Reconstructing the duplication history of a tandem repeat. In: Proceedings of the Seventh International Conference on Intelligent Systems for Molecular Biology, pp. 44–53. AAAI Press (1999)
4. Berard, S., Rivals, E.: Comparison of minisatellites. Journal of Computational Biology 10(3-4), 357–372 (2003)
5. Bertrand, D., Lajoie, M., El-Mabrouk, N.: Inferring ancestral gene orders for a family of tandemly arrayed genes. Journal of Computational Biology 15(8), 1063–1077 (2008)
6. Chauve, C., Doyon, J.P., El-Mabrouk, N.: Gene family evolution by duplication, speciation, and loss. Journal of Computational Biology 15(8), 1043–1062 (2008)
7. Crochemore, M.: An optimal algorithm for computing the repetitions in a word. Inf. Process. Lett. 12(5), 244–250 (1981)
8. Fitch, W.M.: Phylogenies constrained by the crossover process as illustrated by human hemoglobins and a thirteen-cycle, eleven-amino-acid repeat in human apolipoprotein a-i. Genetics 86(3), 623–644 (1977)
9. Gascuel, O., Hendy, M.D., Jean-Marie, A., McLachlan, R.: The combinatorics of tandem duplication trees. Systematic Biology 52(1), 110–118 (2003)
10. Hauth, A.M., Joseph, D.: Beyond tandem repeats: complex pattern structures and distant regions of similarity. In: ISMB, pp. 31–37 (2002)
11. Jeffreys, A.J., Wilson, V., Thein, S.L.: Individual-specific fingerprints of human DNA. Nature 51(2), 71–88 (1980)
12. Kimura, M.: Estimation of evolutionary distances between homologous nucleotide sequences. Proceedings of the National Academy of Sciences 78(1), 454–458 (1981)
13. Lajoie, M., Bertrand, D., El-Mabrouk, N., Gascuel, O.: Duplication and inversion history of a tandemly repeated genes family. Journal of Computational Biology 14(4), 462–478 (2007)
14. Matroud, A.A., Hendy, M.D., Tuffley, C.P.: Ntrfinder: a software tool to find nested tandem repeats. Nucleic Acids Research 40(3), e17 (2012)
15. Rivals, E.: A survey on algorithmic aspects of tandem repeats evolution. Int. J. Found. Comput. Sci. 15(2), 225–257 (2004)
16. Sagot, M.F., Myers, E.W.: Identifying satellites and periodic repetitions in biological sequences. Journal of Computational Biology 5(3), 539–554 (1998)
17. Sammeth, M., Stoye, J.: Comparing tandem repeats with duplications and excisions of variable degree. IEEE/ACM Transactions on Computational Biology and Bioinformatics 3, 395–407 (2006)
18. Stoye, J., Gusfield, D.: Simple and flexible detection of contiguous repeats using a suffix tree. Theor. Comput. Sci. 270(1-2), 843–856 (2002)

RNA Folding Algorithms with G-Quadruplexes

Ronny Lorenz[1], Stephan H. Bernhart[2], Fabian Externbrink[2],
Jing Qin[3], Christian Höner zu Siederdissen[1], Fabian Amman[1],
Ivo L. Hofacker[1,4], and Peter F. Stadler[1,2,3,4,5,6]

[1] Dept. Theoretical Chemistry, Univ. Vienna, Währingerstr. 17, Wien, Austria
[2] Dept. Computer Science, and Interdisciplinary Center for Bioinformatics,
Univ. Leipzig, Härtelstr. 16-18, Leipzig, Germany
[3] MPI Mathematics in the Sciences, Inselstr. 22, Leipzig, Germany
[4] RTH, Univ. Copenhagen, Grønnegårdsvej 3, Frederiksberg C, Denmark
[5] FHI Cell Therapy and Immunology, Perlickstr. 1, Leipzig, Germany
[6] Santa Fe Institute, 1399 Hyde Park Rd., Santa Fe, USA

Abstract. G-quadruplexes are abundant locally stable structural elements in nucleic acids. The combinatorial theory of RNA structures and the dynamic programming algorithms for RNA secondary structure prediction are extended here to incorporate G-quadruplexes using a simple but plausible energy model. With preliminary energy parameters we find that the overwhelming majority of putative quadruplex-forming sequences in the human genome are likely to fold into canonical secondary structures instead.

Keywords: Dynamic programming, RNA folding, `ViennaRNA Package`.

1 Introduction

Guanosine-rich nucleic acid sequences readily fold into four-stranded structures known as G-quadruplexes. DNA quadruplexes are, for instance, an important component of human telomeres [35], they appear to be strongly overrepresented in the promoter regions of diverse organisms, and they can associate with a variety of small molecule ligands, see [23,42] for recent reviews. SNPs in G-quadruplexes, finally, have been implicated as a source variation of gene expression levels [2]. RNA quadruplexes have also been implicated in regulatory functions. Conserved G-quadruplex structures within the 5'-UTR of the human TRF2 mRNA [12] and eukaryotic MT3 matrix metalloproteinases, for example, repress translation [33]. Another well-studied example is the interaction of the RGG box domain fragile X mental retardation protein (FMRP) to a G-quartet-forming region in the human semaphorin 3F (S3F) mRNA [31,4]. A recent review of G-quadruplex-based translation regulation is [8]. A functional RNA G-quadruplex in the 3' UTR was recently described as a translational repressor of the proto-oncogene PIM1 [1]. A mechanistic study of this effect, which seems to be widely used in the cell [19,3], can be found e.g. in [41]. Most recently, G-quadruplexes were also reported in several long non-coding RNAs

M.C.P. de Souto and M.G. Kann (Eds.): BSB 2012, LNBI 7409, pp. 49–60, 2012.
© Springer-Verlag Berlin Heidelberg 2012

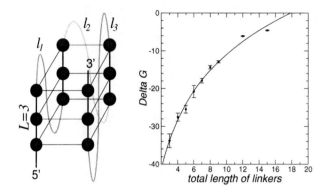

Fig. 1. RNA quadruplexes form parallel arrangements with $L = 2...5$ layers. Folding energies for $L = 3$ depend mostly on the total length ℓ of the linker sequences: the data from ref. [43] fit well to an energy model of the form $\Delta G = a + b \ln \ell$ (solid line).

[21]. G-quadruplexes are potentially of functional importance in the 100 to 9000 nt G-rich telomeric repeat-containing RNAs (TERRAs) [28].

Quadruplex structures consist of stacked associations of G-quartets, i.e., planar assemblies of four Hoogsteen-bonded guanines. As in the case of base pairing, the stability of quadruplexes is derived from π-orbital interactions among stacked quartets. The centrally located cations that are coordinated by the quartets also have a major influence on the stability of quadruplex structures.

DNA quadruplexes are structurally heterogeneous: depending on the glycosidic bond angles there are 16 possible structures and further combinatorial complexity is introduced by the relative orientations of the backbone along the four edges of the stack [37]. RNA quadruplexes, in contrast, appear to be structurally monomorphic forming parallel-stranded conformations (Fig. 1, left) independently of surrounding conditions, i.e., different cations and RNA concentration [45]. Here, we restrict ourselves to the simpler case of RNA quadruplexes.

Bioinformatically, G-quadruplex structures have been investigated mostly as genomic sequence motifs. The `G4P Calculator` searches for four adjacent runs of at least three Gs. With its help a correlation of putative quadruplex forming sequences and certain functional classes of genes was detected [10]. Similarly, `quadparser` [18] recognizes the pattern (1) below. It was used e.g. in [46] to demonstrate the enrichment of quadruplexes in transcriptional regulatory regions. A substantial conservation of such sequence patterns in mammalian promoter regions is reported in [40]. The web service `QGRS Mapper` uses a similar pattern and implements a heuristic scoring system [24], see also [39] for a review. A Bayesian prediction framework based on Gaussian process regression was recently introduced to predict melting temperatures of quadruplex sequences [38].

The formation of RNA quadruplexes necessarily competes with the formation of canonical secondary structures. Hence they cannot be fully understood in isolation. In this contribution we therefore investigate how G-quadruplex structures can be incorporated into RNA secondary structure prediction algorithms.

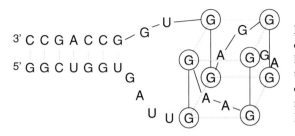

Fig. 2. Structure of the G-quadruplex in a hairpin of human semaphorin 3F RNA that binds the RGG box domain of fragile X mental retardation protein (FMRP). Redrawn based on [31].

2   Energy Model for RNA Quadruplexes

Thermodynamic parameters for RNA quadruplexes can be derived from measurements of UV absorption as a function of temperature [32], analogous to melting curves of secondary structures. While the stability of DNA G-quadruplexes strongly depends on the arrangement of loops [7,44] this does not appear to be the case for RNA. RNA not only forms mostly parallel-stranded stacks for G-quartets but their stability also exhibits a rather simple dependence of the loop length [43]. In further contrast to DNA [14], they appear to be less dependent on the nucleotide sequence itself.

A G-quadruplex with $2 \leq L \leq 5$ stacked G-quartets and three linkers of length $l_1, l_2, l_3 \geq 1$ has the form

$$G_L N_{l_1} G_L N_{l_2} G_L N_{l_3} G_L \tag{1}$$

It is commonly assumed that $1 \leq l_i \leq 7$ [38], although *in vitro* data for DNA suggest that longer linkers are possible [15]. For $L = 2$, the existence of quadruplexes with $1 \leq \ell_i \leq 2$ was reported [26]. For $L = 3$ detailed thermodynamic data are available only for the 27 cases $1 \leq l_1, l_2, l_3 \leq 3$ and for some longer symmetric linkers $l_1 = l_2 = l_3$ [43], see Figure 1b. To our knowledge, no comprehensive data are available for $L \geq 4$. It appears reasonable to assume that the stacking energies are additive. The energetic effect of the linkers appears to be well described in terms of the total linker length ℓ [43]. As shown in Figure 1b the free energy depends approximately logarithmically on ℓ. In this contribution we are mostly concerned with the algorithmic issues of including G-quadruplexes into thermodynamic folding programs. In particular we ignore here the strong dependence of quadruplex stability on the potassium concentration, see e.g. [22]. We thus resort to the simplified energy function

$$E[L, \ell] = a(L - 1)g_0 + b \ln(\ell - 2) \tag{2}$$

with parameters $a = -18$ kcal/mol and $b = 12$ kcal/mol if the pattern (1) is matched, and $E = \infty$ otherwise.

G-quadruplex structures can be located within loops of more complex secondary structures. Fig. 2, for instance, shows the $L = 2$, $l_1 = l_2 = l_3 = 2$ quadruplex in a hairpin of the semaphorin 3F RNA [31]. It seems natural to treat G-quadruplexes inside multiloops similar to their branching helices: each unpaired base incurs a penalty a and each G-quadruplex within a loop is associated with an additional "loop strain" b. For the interior-loop case of Fig. 2, only stabilizing mismatch contributions of the enclosing pair and a penalty for the stretches of unpaired bases are used. Sterical considerations for this case suggest that a G-quadruplex is flanked by a stretch of at least three unpaired nucleotides or has at least one unpaired nucleotide on either side.

3   Combinatorics of Structures with Quadruplexes

RNA secondary structures consist of mutually non-crossing base pairs and unpaired positions. Thus they can be represented as strings composed of matching parentheses (base pairs) and dots. This "dot-parenthesis" notation is used by the `ViennaRNA Package` [27]. G-quadruplexes constitute an extra type of structural element. The semaphorin hairpin, Fig. 2, can therefore be written as

$$
\begin{array}{l}
\text{GGCUGGUGAUUGGAAGGGAGGGAGGUGGCCAGCC} \\
\text{(((((((....++..++..++..++..)))))))}
\end{array} \tag{3}
$$

using the symbol + to mark the bases involved in G-quartets. This string representation uniquely identifies all G-quartets since the first run of + symbols determines L for the 5'-most quadruplex, thus determining the next three G-stacks which are separated by at least one '.' and must have the same length. It follows immediately that the number of secondary structures with G-quadruplexes is still smaller than 4^n, an observation that is important for the evolvability of RNAs [36]. In order to get a tighter bound on the number of structures we use here, for the sake of presentation, a simplified model in which we omit the restrictions of a minimal size of a hairpin loop and allow quadruplexes with any value of $L \geq 2$ and $l_i \geq 1$.

Let $\mathbf{g}_n$ denote the number of secondary structures with G quadruplexes on a sequence of length n. The corresponding generating function is $\mathbf{G}(x) = \sum_{n \geq 0} \mathbf{g}_n x^n$. Similarly, let $\mathbf{q}_n$ be the number of quadruplexes on length n. As derived in the supplement, its generating function is $\mathbf{Q}(x) = \sum_{n \geq 0} \mathbf{q}_n x^n = x^{11}(1-x)^{-3}(1-x^4)^{-1}$. The basic idea is now to consider a structure consisting of b base pairs, u unpaired bases and k quadruplexes. Then there are $\binom{2b+k}{k}$ ways to insert k quadruplexes into each of the $C_b = \frac{1}{b+1}\binom{2b}{b}$ possible arrangements of b matching pairs of parentheses. Into each of these arrangements we can insert u unpaired bases in $\binom{2b+k+u}{u}$ different ways. Thus we have

$$
\mathbf{G}(x) = \sum_k \sum_b \sum_u \frac{1}{b+1}\binom{2b}{b}\binom{2b+k}{k}\binom{2b+k+u}{u} x^{2b+u}\mathbf{Q}(x)^k
$$

$$
= \frac{2}{1 - x - \mathbf{Q}(x) + \sqrt{(1 - 3x - \mathbf{Q}(x))(1 + x - \mathbf{Q}(x))}}
\tag{4}
$$

Following [11] we find that the coefficients of $\mathbf{G}(x)$ are asymptotically given by $\mathbf{g}_n \sim k_0 n^{-3/2} \gamma^n$, where k_0 is a positive constant and $\gamma \approx 3.00005$. A more

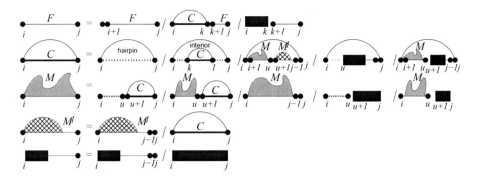

Fig. 3. Extension of recursions of the `ViennaRNA Package` to accomodate G-quadruplexes. This grammar treats G-quadruplexes with multi-loop like energies also in an interior-loop-like context.

detailed model accounting for minimal stack and loop lengths is analyzed in the Supplemental Material. It yields $\gamma \approx 2.2903$ if isolated base pairs are allowed, and $\gamma \approx 1.8643$ for canonical secondary structures.

4   RNA Folding Algorithms

Energy Minimization. Dynamic programming algorithms for secondary structure prediction are based on a simple recursive decomposition: any feasible structure on the interval $[i, j]$ has the first base either unpaired or paired with a position k satisfying $i < k \leq j$. The condition that base pairs do not cross implies that the intervals $[i+1, k-1]$ and $[k+1, j]$ form self-contained structures whose energies can be evaluated independent of each other. In conjunction with the standard energy model [29], which distinguishes hairpin loops, interior loops (including stacked base pairs), and multi-loops, this leads to the recursions diagrammatically represented in Fig. 3 (ignoring the cases involving black blocks). This algorithmic approach was pioneered e.g. in [47,30] and is also used in the `ViennaRNA Package` [27].

G-quadruplexes form closed structural elements on well-defined sequence intervals. Thus they can be treated just like substructures enclosed by a base pair, so that the additional ingredients in the folding algorithms are the energies G_{ij} (free energy of the most stable quadruplex so that the pattern (1) matches exactly the interval $[i, j]$) and the partition functions Z_{ij}^G (defined as the sum of the Boltzmann factors of all distinct quadruplexes on the interval $[i, j]$). As a consequence of (1) we have $G_{ij} < \infty$ and $Z_{ij}^G > 0$ only if $|j - i| < 4L_{\max} + \ell_{\max}$. All possible quadruplexes on the interval $[i, j]$ can be determined and evaluated in $\mathcal{O}(L_{\max}^2 \ell_{\max}^2)$ time so that these arrays can be precomputed in $\mathcal{O}(n(L_{\max} + \ell_{\max})L_{\max}^2 \ell_{\max}^2)$, i.e., in linear time.

The standard recursions for RNA secondary structure prediction can now be extended by extra terms for quadruplexes, see Fig. 3. The simplest strategy would be to add G-quadruplexes as an additional type of base-pair enclosed structures. This would amount to using standard interior loop parameters also for cases such as Fig. 2. Hence we use the somewhat more elaborate grammar of Fig. 3, which introduces the quadruplexes in the form of additional cases into the multi-loop decomposition. An advantage of this method is that one can use different parameter values to penalize the inclusion of quadruplexes and helical components into a multiloop. Clearly the grammar is still unambiguous, i.e., every structure has an unique parse. Thus it can be used directly to compute partition functions.

Base Pairing Probabilities. A straightforward generalization of McCaskill's algorithm can be used to compute the probabilities P_{ij} of all possible base pairs (i, j). The probability P_{ij}^G of finding a G-quadruplex delimited by positions i and j then can be written as

$$P_{ij}^G = \frac{Z_{1,i-1} Z_{ij}^G Z_{j+1,n}}{Z} + \sum_{\substack{k<i-1 \\ l>j+1}} P_{kl} \mathbb{P}\left\{\text{quadruplex}[i, j]\big|(k, l)\right\} \qquad (5)$$

The conditional probabilities $\mathbb{P}\{\dots\}$ in turn are composed of the four individual cases depending on the placement of the components of the generalized multiloop enclosed by (k, l) relative to the interval $[i, j]$:

$$(6)$$

This decomposition translates to the recursion $\mathbb{P}\left\{\text{quadruplex}[i, j]\big|(k, l)\right\} =$

$$\frac{Z_{k+1,i-1}^M Z_{ij}^G Z_{j+1,l-1}^M}{Z_{kl}^B} + \frac{Z_{k+1,i-1}^M Z_{ij}^G \hat{b}^{l-j-1}}{Z_{kl}^B} + \frac{\hat{b}^{i-k-1} Z_{ij}^G Z_{j+1,l-1}^M}{Z_{kl}^B} + \frac{\hat{b}^{i-k-1} Z_{ij}^G \hat{b}^{l-j-1}}{Z_{kl}^B}$$

where $\hat{b} = \exp(-b/RT)$. From the P_{ij}^G it is straightforward to compute the probability of a particular quadruplex as

$$p([i, L, l_1, l_2, j]) = \frac{\exp(-E[L, \ell])}{Z_{ij}^G} P_{ij}^G \qquad (7)$$

where $l_3 = j - i + 1 - 4L - l_1 - l_2$. Summing up the probabilities of all quadruplexes that contain a particular contact $i' : j'$ of two guanosines in a layer finally yield the probability of the G:G contact $i' : j'$.

Fig. 4 shows an example of the graphical output of RNAfold. In the minimum energy case we use a very simple modification of the standard layout [6] treating each quadruplex like a local hairpin structure, explicitly indicating the G-G pairs. Quadruplexes are shown in addition to the individual G-G pairs as shaded triangles in the base pair probability dot plots. From the base pairing probabilities we also compute MEA [9] and centroid structures.

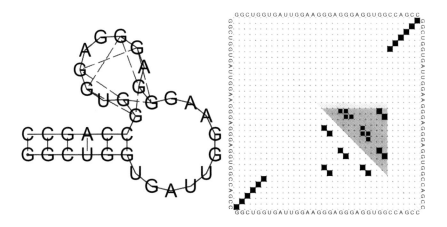

Fig. 4. Representation of minimum free energy structure (l.h.s.) and base pairing probability matrix (r.h.s.) of the semaphorin hairpin (see Fig. 2) respectively

By definition the centroid structure X minimizes the expected base pair distance to the other structures within the Boltzmann-weighted ensemble. In the absence of G-quadruplexes X consists of all base pairs (i, j) with $p_{ij} > 1/2$. A certain ambiguity arises depending on whether X is interpreted as a list of base pairs that may contain incomplete quadruplexes, or whether quadruplexes are treated as units. Here, we insert a quadruplex if $P_{ij}^G > 0.5$, and represent it by the most stable quadruplex with endpoints i and j. The same representation is used for MEA structures where we extend the maximized expected accuracy to $EA = \sum_{(i,j) \in S} 2\gamma(P_{i,j} + P_{ij}^G) + \sum_i P_i^u$ with $P_i^u = 1 - \sum_j P_{ij} - \sum_{k \leq i \leq l} P_{kl}^G$, accordingly.

Consensus Structures can be readily obtained for a given multiple sequence alignment. The idea is to apply the dynamic programming recursions to alignment columns. The energy contributions are determined as the average of the corresponding contributions to the individual sequences [16]. In addition small contributions are added to favor pairs of columns with consistent (e.g. GC→GU) and compensatory mutations (AU→GC) since these provide direct evidence for selection acting to preserve base pairing. Similarly, penalties are added if one or a few sequences cannot form a base pair. We refer to [5] for details of the scoring model implemented in RNAalifold. Here, we extend it by a simple system of penalties for mutations that disrupt quadruplexes. Non-G nucleotides incur an energy E' in the outer layers of the quadruplex and $2E'$ in the inner layers as they affect one or two stacking interactions, respectively. An example of a consensus structure prediction is shown in Fig. 5.

Implementation Details. The implementation of G-quadruplex folding in RNAfold and RNAalifold essentially follows the extended grammar shown in Fig. 3, distinguishing the energy contribution of unpaired bases in the external loop from those enclosed by base pairs. The energies of all possible G-quadruplexes are

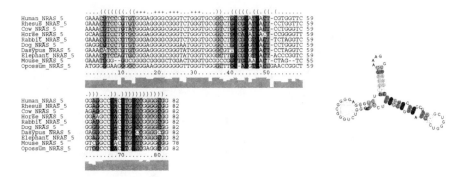

Fig. 5. Consensus structure of the 5'-most part of the 5'UTR of the NRAS mRNA, exhibiting a conserved G-quadruplex with $L = 3$ that modulates translation of the *NRAS* proto-oncogene [25]. Colors indicate the number (red 1, ochre 2, green 3) of different types of basepairs in a pair of alignment columns, unsaturated colors indicate basepairs that cannot be formed by 1 or 2 sequences. Substitutions in stem regions are indicated by circles in the secondary structure drawing.

pre-computed, storing the energy of the most stable quadruplex for each pair of endpoints in the triangular matrix G. As this matrix will be very sparse for most inputs, a sparse matrix optimization is possible, but not yet implemented. In the backtracing part we re-enumerate quadruplexes with given endpoints whenever necessary. Base pairing probabilities are computed as outlined above. Since there cannot be a conflict with canonical base pairs, we store P_{ij}^G as part of the base pairing probability matrix. The probabilities of individual G-G contacts are computed by enumeration as a post-processing step. We also adapted the `RNAeval` and `RNAplot` programs so that sequence/structure pairs can be parsed and re-evaluated according to the extended grammar.

5   Evaluation

Runtime Performance. The runtime of `RNAfold` with the extended grammar of Fig. 3 was compared to the implementation of the standard model. For both, energy minimization and partition function, virtually no difference was observed. For short sequences of about 200 nt the additional pre-processing steps incur a minor but negligible runtime overhead.

Occurrence and Stability of G-quadruplexes in Genomes. Sequence motifs of the form (1) that can in principle form quadruplex structures are very abundant in most genomes, see e.g. [10,18,46]. The number of putative quadruplex-forming sequences is even slightly larger than expected from random sequences with the same mono- or dinucleotide distributions, Fig. 6. The overwhelming majority of these quadruplex candidates, however, is unstable compared to canonical secondary structures that use some or all of Gs in canonical base pairs. We observe that less than 2% of the putative quadruplexes are thermodynamically

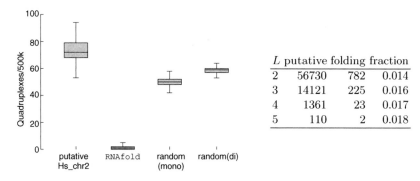

Fig. 6. Abundance and stability of putative G-quadruplexes. L.h.s.: Box plot showing the number of potential G-quadruplexes in human chromosome 2 within sliding windows of 500 000 nucleotides. For comparison, the same information for a random sequence with the same mono- or di-nucleotide composition than chr. 2 is presented as well. Both, the mono- and di-nucleotide distribution have been generated from chromosome 2. `RNAfold` denotes the number of putative G-quadruplexes stable enough to occur in a predicted structure of 100 nucleotides up- and downstream of the putative G-quadruplex (with median=1, interquartile range=0–2). R.h.s.: fraction of stable quadruplexes as function of L. for human chromosome 2.

stable. Interestingly, this effect is nearly independent of the number of layers (L). This data is preliminary, since it is based on energy parameters that have been fitted on a rather limited amount of empirical evidence and do not include the important issues arising from the strong dependence of quadruplex stability of cation concentrations. Furthermore, it reflects the occurrance of putative G-quadruplexes on the human chromosome 2 only. More comprehensive and accurate parameters as well as a local folding algorithm that extends `RNALfold` [17] will be subject of a forthcoming study. Several experimentally know RNA quadruplexes are predicted by the current version, including the semphorin hairpin of Fig. 4 and the quadruplex in human telomerase RNA [13] discussed in the Supplemental Material.

6   Discussion

We have shown in this contribution that structural elements such as G-quadruplexes that correspond to uninterrupted sequence intervals can be included in a rather straightforward way into the standard dynamic programming recursions – provided a corresponding extension of the energy model can be devised. The G-quadruplex-aware programs are currently available as a separate branch (version number with the suffix "g") of the `ViennaRNA Package` using a very simple energy function for the quadruplexes that reproduces the few available experimental data at least semi-quantitatively. Following further optimization of the code the algorithmic extensions will be integrated in the main version of the package in the near future. The extensions in Fig. 3 can also be applied to local folding algorithms such as `RNALfold` and `RNAplfold` or the exhaustive enumeration of suboptimal structures in `RNAsubopt`. This is ongoing work, as is a comprehensive set of tools for genome-wide scans for putative G-quadruplexes.

It is less obvious how to handle quadruplexes in RNA-RNA interactions since our recursions consider local G-quadruplexes only. At least it is clear that they can be included in all those parts of the structure that are not involved in intermolecular contacts. Some quadruplex structures, however, are formed *in trans*. The binding of G-rich small RNAs to G-rich regions in reporter mRNAs leads to the formation of an intermolecular RNA G-quadruplex that in turn can inhibit translation in living cells [20]. One can use `RNAup` [34] to compute the probabilities $p^{(1)}$ and $p^{(2)}$ that the G-rich regions are unpaired. From these, we obtain the free energies $G^{(i)} = -RT \ln p^{(i)}$ to make the binding site accessible. It remains to compute the interaction energy itself.

The main problem for practical applications of quadruplex-aware RNA folding tools is our limited knowledge of the energy function in particular for $L \neq 3$ and for asymmetric linkers. Even with the crude energy function employed here it becomes clear that the overwhelming majority of putative genomic quadruplex sequences will fold into a canonical secondary structure rather than G-quadruplex structures.

Supplemental Material and Source Code. Available from http://www.bioinf.uni-leipzig.de/publications/supplements/12-006 and www.tbi.univie.ac.at/ ronny/programs/.

Acknowledgements. This work was supported in part by the German Research Foundation (STA 850/7-2, under the auspicies of SPP-1258 "Sensory and Regulatory RNAs in Prokaryotes"), the Austrian GEN-AU projects "regulatory non coding RNA", "Bioinformatics Integration Network III" and the Austrian FWF project "SFB F43 RNA regulation of the transcriptome".

References

1. Arora, A., Suess, B.: An RNA G-quadruplex in the 3' UTR of the proto-oncogene PIM1 represses translation. RNA Biology 8, 802–805 (2011)
2. Baral, A., Kumar, P., Halder, R., Mani, P., Yadav, V.K., Singh, A., Das, S.K., Chowdhury, S.: Quadruplex-single nucleotide polymorphisms (Quad-SNP) influence gene expression difference among individuals. Nucleic Acids Res. (2012)
3. Beaudoin, J.D., Perreault, J.P.: 5'-UTR G-quadruplex structures acting as translational repressors. Nucleic Acids Res. 38, 7022–7036 (2010)
4. Bensaid, M., Melko, M., Bechara, E.G., Davidovic, L., Berretta, A., Catania, M.V., Gecz, J., Lalli, E., Bardoni, B.: FRAXE-associated mental retardation protein (FMR2) is an RNA-binding protein with high affinity for G-quartet RNA forming structure. Nucleic Acids Res. 37, 1269–1279 (2009)
5. Bernhart, S.H., Hofacker, I.L., Will, S., Gruber, A.R., Stadler, P.F.: `RNAalifold`: improved consensus structure prediction for RNA alignments. BMC Bioinformatics 9, 474 (2008)

6. Bruccoleri, R.E., Heinrich, G.: An improved algorithm for nucleic acid secondary structure display. Computer Appl. Biosci. 4, 167–173 (1988)
7. Bugaut, A., Balasubramanian, S.: A sequence-independent study of the influence of short loop lengths on the stability and topology of intramolecular DNA G-quadruplexes. Biochemistry 47, 689–697 (2008)
8. Bugaut, A., Balasubramanian, S.: 5'-UTR RNA G-quadruplexes: translation regulation and targeting. Nucleic Acids Res. (2012), doi: 10.1093/nar/gks068
9. Do, C.B., Woods, D.A., Batzoglou, S.: CONTRAfold: RNA secondary structure prediction without physics-based models. Bioinformatics 22(14), e90–e98 (2006)
10. Eddy, J., Maizels, N.: Gene function correlates with potential for G4 DNA formation in the human genome. Nucleic Acids Res. 34, 3887–3896 (2006)
11. Flajolet, P., Sedgewick, R.: Analytic Combinatorics. Cambridge University Press, New York (2009)
12. Gomez, D., Guédin, A., Mergny, J.L., Salles, B., Riou, J.F., Teulade-Fichou, M.P., Calsou, P.: A G-quadruplex structure within the 5'-UTR of TRF2 mRNA represses translation in human cells. Nucleic Acids Res. 38, 7187–7198 (2010)
13. Gros, J., Guédin, A., Mergny, J.L., Lacroix, L.: G-Quadruplex formation interferes with P1 helix formation in the RNA component of telomerase hTERC. ChemBioChem 9, 2075–2079 (2008)
14. Guédin, A., De Cian, A., Gros, J., Lacroix, L., Mergny, J.L.: Sequence effects in single-base loops for quadruplexes. Biochimie 90, 686–696 (2008)
15. Guédin, A., Gros, J., Patrizia, A., Mergny, J.L.: How long is too long? Effects of loop size on G-quadruplex stability. Nucleic Acids Res. 38, 7858–7868 (2010)
16. Hofacker, I.L., Fekete, M., Stadler, P.F.: Secondary structure prediction for aligned RNA sequences. J. Mol. Biol. 319, 1059–1066 (2002)
17. Hofacker, I.L., Priwitzer, B., Stadler, P.F.: Prediction of locally stable RNA secondary structures for genome-wide surveys. Bioinformatics 20, 191–198 (2004)
18. Huppert, J.L., Balasubramanian, S.: Prevalence of quadruplexes in the human genome. Nucleic Acids Res. 33, 2908–2916 (2005)
19. Huppert, J.L., Bugaut, A., Kumari, S., Balasubramanian, S.: G-quadruplexes: the beginning and end of UTRs. Nucleic Acids Res. 36, 6260–6268 (2008)
20. Ito, K., Go, S., Komiyama, M., Xu, Y.: Inhibition of translation by small RNA-stabilized mRNA structures in human cells. J. Am. Chem. Soc. 133, 19153–19159 (2011)
21. Jayaraj, G.G., Pandey, S., Scaria, V., Maiti, S.: Potential G-quadruplexes in the human long non-coding transcriptome. RNA Biolog. 9, 81–86 (2012)
22. Joachimi, A., Benz, A., Hartig, J.S.: A comparison of DNA and RNA quadruplex structures and stabilities. Bioorg. Med. Chem. 17, 6811–6815 (2009)
23. Johnson, J.E., Smith, J.S., Kozak, M.L., Johnson, F.B.: *In vivo veritas*: using yeast to probe the biological functions of G-quadruplexes. Biochimie 90, 1250–1263 (2008)
24. Kikin, O., D'Antonio, L., Bagga, P.S.: QGRS mapper: a web-based server for predicting G-quadruplexes in nucleotide sequences. Nucleic Acids Res. 34, W676–W682 (2006)
25. Kumari, S., Bugaut, A., Huppert, J.L., Balasubramanian, S.: An RNA G-quadruplex in the 5'UTR of the NRAS proto-oncogene modulates translation. Nat. Chem. Biol. 3, 218–221 (2007)
26. Lauhon, C.T., Szostak, J.W.: RNA aptamers that bind flavin and nicotinamide redox cofactors. J. Am. Chem. Soc. 117, 1246–1257 (1995)
27. Lorenz, R., Bernhart, S.H., Höner zu Siederissen, C., Tafer, H., Flamm, C., Stadler, P.F., Hofacker, I.L.: ViennaRNA Package 2.0. Alg. Mol. Biol. 6, 26 (2011)

28. Luke, B., Lingner, J.: TERRA: telomeric repeat-containing RNA. EMBO J. 28, 2503–2510 (2009)

29. Mathews, D.H., Disney, M.D., Childs, J.L., Schroeder, S.J., Zuker, M., Turner, D.H.: Incorporating chemical modification constraints into a dynamic programming algorithm for prediction of RNA secondary structure. Proc. Natl. Acad. Sci. USA 101, 7287–7292 (2004)

30. McCaskill, J.S.: The equilibrium partition function and base pair binding probabilities for RNA secondary structure. Biopolymers 29, 1105–1119 (1990)

31. Menon, L., Mihailescu, M.R.: Interactions of the G quartet forming semaphorin 3F RNA with the RGG box domain of the fragile X protein family. Nucleic Acids Res. 35, 5379–5392 (2007)

32. Mergny, J.L., Lacroix, L.: UV melting of G-quadruplexes. Curr. Protoc. Nucleic Acid Chem. Unit 17.1 (2009)

33. Morris, M.J., Basu, S.: An unusually stable G-quadruplex within the 5'-UTR of the MT3 matrix metalloproteinase mRNA represses translation in eukaryotic cells. Biochemistry 48, 5313–5319 (2009)

34. Mückstein, U., Tafer, H., Hackermüller, J., Bernhard, S.B., Stadler, P.F., Hofacker, I.L.: Thermodynamics of RNA-RNA binding. Bioinformatics 22, 1177–1182 (2006)

35. Paeschke, K., Simonsson, T., Postberg, J., Rhodes, D., Lipps, H.J.: Telomere end-binding proteins control the formation of G-quadruplex DNA structures *in vivo*. Nature Struct. Mol. Biol. 12, 847–854 (2005)

36. Schuster, P., Fontana, W., Stadler, P.F., Hofacker, I.L.: From sequences to shapes and back: A case study in RNA secondary structures. Proc. Roy. Soc. Lond. B 255, 279–284 (1994)

37. Webba da Silva, M.: Geometric formalism for DNA quadruplex folding. Chemistry 13, 9738–9745 (2007)

38. Stegle, O., Payet, L., Mergny, J.L., MacKay, D.J.C., Huppert, J.L.: Predicting and understanding the stability of G-quadruplexes. Bioinformatics 25, i374–i382 (2009)

39. Todd, A.K.: Bioinformatics approaches to quadruplex sequence location. Methods 43, 246–251 (2007)

40. Verma, A., Halder, K., Halder, R., Yadav, V.K., Rawal, P., Thakur, R.K., Mohd, F., Sharma, A., Chowdhury, S.: G-quadruplex DNA motifs as conserved cis-regulatory elements. J. Med. Chem. 51, 5641–5649 (2008)

41. Wieland, M., Hartig, J.S.: RNA quadruplex-based modulation of gene expression. Chem. Biol. 14, 757–763 (2007)

42. Wong, H.M., Payet, L., Huppert, J.L.: Function and targeting of G-quadruplexes. Curr. Opin. Mol. Ther. 11, 146–155 (2009)

43. Zhang, A.Y., Bugaut, A., Balasubramanian, S.: A sequence-independent analysis of the loop length dependence of intramolecular RNA G-quadruplex stability and topology. Biochemistry 50, 7251–7258 (2011)

44. Zhang, D.H., Fujimoto, T., Saxena, S., Yu, H.Q., Miyoshi, D., Sugimoto, N.: Monomorphic RNA G-quadruplex and polymorphic DNA G-quadruplex structures responding to cellular environmental factors. Biochemistry 49, 4554–4563 (2010)

45. Zhang, D.H., Zhi, G.Y.: Structure monomorphism of RNA G-quadruplex that is independent of surrounding condition. J. Biotechnol. 150, 6–10 (2010)

46. Zhao, Y., Du, Z., Li, N.: Extensive selection for the enrichment of G4 DNA motifs in transcriptional regulatory regions of warm blooded animals. FEBS Letters 581, 1951–1956 (2007)

47. Zuker, M., Stiegler, P.: Optimal computer folding of large RNA sequences using thermodynamics and auxiliary information. Nucleic Acids Res. 9, 133–148 (1981)

Molecular Dynamics for Simulating the Protein Folding Process Using the 3D AB Off-Lattice Model

César Manuel Vargas Benítez and Heitor Silvério Lopes

Bioinformatics Laboratory, Federal University of Technology - Paraná
Av. 7 de setembro, 3165 80230-901, Curitiba (PR), Brazil
cesarvargasb@gmail.com, hslopes@utfpr.edu.br

Abstract. To the best of our knowledge, this paper presents the first application of Molecular Dynamics to the Protein Folding Problem using the 3D AB model of proteins. Protein folding pathways are also presented and discussed. This work also offered new reference values for five benchmark sequences. Future works will investigate parallel versions of the presented approach and more experiments to create new bechmarks.

Keywords: Protein Folding, 3D AB model, Molecular Dynamics.

1 Introduction

Proteins are the basic structures of all living beings. Finding the proteins that make up an organism and understanding their function is the foundation of Molecular Biology [11]. They are polymers composed by a chain of amino acids (also called residues) that are linked together by means of peptide bonds, and are synthesized in the ribosome of cells following a template given by the messenger RNA (mRNA). Under physiological conditions every protein folds into a unique three-dimensional structure that determines their specific biological function. This structure is called the native tertiary structure and it depends on its amino acids sequence [1].

2 The Protein Folding Problem

In vitro experiments carried out by Christian Anfinsen and colleagues [1] show that proteins can be denatured by modifications in the environment where they are. Most proteins can be denatured by temperature and pH variations, affecting weak interactions between residues (i.e.: hydrogen bonds). During the denaturation process, proteins lose their native shape and, consequently, their function. Anfinsen showed that denatured (misfolded or unfold) proteins can refold into their native conformation. However, the spontaneous refolding only occurs in single-domain proteins. In the worst case, such misfolded proteins can be completely inactive or even harmful to the organism. Several diseases such

M.C.P. de Souto and M.G. Kann (Eds.): BSB 2012, LNBI 7409, pp. 61–72, 2012.
© Springer-Verlag Berlin Heidelberg 2012

as cystic fibrosis, Alzheimer's disease, Huntington's disease and some types of cancer are believed to result from the aggregation of misfolded proteins.

One of the most important and challenging problems in Molecular Biology with applications, such as drug design, is to obtain a better understanding of the Protein Folding Problem (PFP). The PFP also has great practical importance in this era of genomic sequencing. For instance, thanks to the several genome sequencing projects being conducted in the world, a large number of new proteins have been discovered. However, only a small number of such proteins have their three-dimensional structure known. The UniProtKB/TrEMBL[1] repository of protein sequences has currently around 20 million records (as in march/2012), and the Protein Data Bank (PDB[2]) has the structure of only 79,850 proteins. This fact is due to the cost and difficulty of unveiling the structure of proteins, from the biochemical point of view. Over more than five decades the protein folding field has evolved. Computer Science has an important role here, proposing models and computation approaches for studying the PFP. In contemporary Computational Biology, there are two protein folding problems. The first problem is to predict the protein structure (conformation) from sequence (primary structure), and the second one is to predict protein folding pathways, which consists in determining the folding sequence of events which lead from the primary structure of a protein (its linear sequence of amino acids) to its native structure. There are many computational methods to deal with the folding problem. However, the Molecular Dynamics (MD) approach (including all its variations) is the only computational methodology that really provides a time-dependent analysis of a system in Molecular Biology and, consequently, can be employed to solve the second PFP [16].

Ideally, both the protein and the solvent should be represented at atomistic level because this approach is the closest to experiments [19]. However, the simulation of computational models that take into account all the atoms of a protein is frequently unfeasible due to the multidimensionality of the system ($> 10^4$ degrees of freedom) [16], even with the most powerful computational resources (in nature, proteins can rapidly and reliably find their way into well-defined folded configurations). Generally, atomistic simulations of real-size proteins are usually limited to unfolding the native conformation of the proteins followed by refolding [19]. The dimensionality of a system containing the protein and the solvent can be reduced when the solvent is treated implicitly and a reduced coarse-grained model of proteins is used. In this scenery, several reduced (mesoscopic) models have been proposed. Although such reduced models are not realistic, their simulation can show some characteristics of real proteins. The success of reduced representations in reproducing several aspects of the folding process is due to the fact that this process has generally evolved to satisfy the principle of minimal frustration [3]. Computational studies of reduced models have provided several valuable insights into the folding process.

---

[1] Available in: http://www.ebi.ac.uk/uniprot/
[2] Available in: http://www.rcsb.org/

Whereas the protein structure prediction problem is widely acknowledged as an open problem, the pathway prediction problem has received little attention. It is important to note that the ability to predict the folding pathways can improve methods for predicting the native structure of proteins.

The total number of possible conformations of a protein is huge and it would take an astronomical length of time to find the native conformation by means of exhaustive search of all conformational space [13]. Nowadays, it is known that the folding process does not include mandatory steps between unfolded and folded states, but a search of many accessible conformations [13]. One approach to enumerate folding pathways is to start with an unfolded protein and consider the various possibilites for the protein to fold. The protein folds from a denatured conformation with a high free energy to its native conformation, following an energy landscape [10]. It is important to know that the free energy barrier between the native state and the multiple denature conformations is huge.

The simplest computational model for the PFP problem is known as Hydrophobic-Polar (HP) model, both in two (2D-HP) and three (3D-HP) dimensions [8]. Although simple, the computational approach for searching a solution for the PFP using HP models was proved to be NP-complete [7].

3   The AB Off-Lattice Model

The AB off-lattice model was introduced by [22] to represent protein structures. In this model each residue is represented by a single interaction site located at the Cα position. These sites are linked by rigid unit-length bonds ($\hat{b}_i$) to form the protein structure.

The three-dimensional structure of an N-length protein is specified by the $N-1$ bond vectors $\hat{b}_i$, $N-2$ bond angles τ_i and $N-3$ torsional angles α_i, as shown in Figure 1.

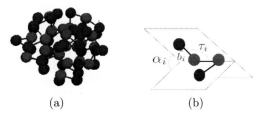

(a)   (b)

Fig. 1. Example of a hypothetic protein structure (a) and Definition of $\hat{b}_i$, τ_i and α_i (b, Adapted from [12]). Blue balls represent the polar residues and Red ones represent the hydrophobic residues. The backbone and the connections between elements are shown in black lines.

The 20 proteinogenic amino acids are classified into two classes, according to their affinity to water (hydrophobicity): 'A' (hydrophobic) and 'B' (hydrophilic or polar). This model do not describe the solvent molecules. However, solvent effects such as the formation of the hydrophobic core are taken into account through interactions between residues, according to their hydrophobicity (species-dependent global interactions).

When a protein is folded into its native conformation, the hydrophobic amino acids tend to pack inside the protein, in such a way to get protected from the solvent by an aggregation of polar amino acids that are positioned outwards. Interactions between amino acids take place and the energy of the conformation tends to decrease. Conversely, the conformation tends to converge to its native state, in accordance with the Anfinsen's thermodynamic hypothesis [1].

The energy function of a folding is given by [12]:

$$E(\hat{b_i}; \sigma) = E_{Angles} + E_{torsion} + E_{LJ} = -k_1 \sum_{i=1}^{N-2} \hat{b_i} \cdot \hat{b_{i+1}}$$

$$-k_2 \sum_{i=1}^{N-3} \hat{b_i} \cdot \hat{b_{i+2}} + \sum_{i=1}^{N-2} \sum_{j=i+2}^{N} 4\varepsilon(\sigma_i, \sigma_j)(r_{ij}^{-12} - r_{ij}^{-6}) \qquad (1)$$

where

$\sigma = \sigma_0, ..., \sigma_N$ form a binary string that represent the protein sequence.
E_{Angles} and $E_{torsion}$ are the energies from bond angle and torsional forces, respectively. Where $\hat{b_i}$ represents the ith bond that joins the $i - 1$th and the ith residues, and is represented by the vector $\hat{b_i} = \boldsymbol{r}_i - \boldsymbol{r}_{i-1}$, and $k1 = -1$; $k2 = +1/2$.

The species-dependent global interactions are given by the Lennard-Jones potencial (E_{LJ}); for pairs of ith and jth residues separated by a distance of r_{ij}. Where $\varepsilon(\sigma_i, \sigma_j)$ is chosen to favor the formation of the hydrophobic core ('A' residues). Thus, $\varepsilon(\sigma_i, \sigma_j)$ is 1 for AA interactions and 1/2 for BB/AB interactions.

4   Molecular Dynamics

Molecular Dynamics (MD) is a computational simulation of physical movements of particles (atoms or molecules). The theoretical basis for MD embodies many of the important results produced by the great names of analytical mechanics – Newton, Euler, Hamilton and Lagrange. The basic form of MD involves little more than Newton's second law [20]. The idea of MD is to generate the trajectory of a system with N particles through numerically integration of the classical equations of motion.

MD is a deterministic approach, differently from Monte Carlo simulations that are stochastic.Thus, a MD simulation will always generate the same trajectory in phase space from the same initial condition.

The explanation of the implemented Molecular Dynamics is given in section 5.

5   Methodology

This section describes in detail the implementation of the Molecular Dynamics algorithm for the PFP using the 3D AB off-lattice model of proteins.

Algorithm 1 shows the pseudo-code of the Molecular Dynamics algorithm.

Algorithm 1. Molecular Dynamics pseudo-code

1: **Start**
2: Set the initial conditions: positions $r_i(t_0)$, velocities $v_i(t_0)$ and accelerations $a_i(t_0)$
3: **while** $t < t_{max}$ **do**
4:     Compute forces on all particles
5:     Integrate equations of motion
6:     Perform ensemble control
7:     Compute geometric constraints
8:     $t \leftarrow t + \delta t$
9: **end while**
10: Compute the desired physical quantities
11: **End**

- **Set the initial conditions**: in this step, initial positions, velocities and accelerations are assigned. An initial unfolded or partially folded conformation is randomly generated. To represent the position of the amino acids, three-dimensional Cartesian coordinates are defined by a vector (x_i, y_i, z_i). The first amino acid of the primary structure is positioned at the origin and next amino acids are positioned at Cartesian coordinates relative to its predecessor and obtained from randomly spherical coordinates, as follows, where $\theta \in [0, 2\pi], \phi \in [0, \pi]$ and $(x_i, y_i, z_i) = (x_{i-1} + r_{ij} * sin\phi * cos\theta, y_{i-1} + r_{ij} * sin\phi * sin\theta, z_{i-1} + r_{ij} * cos\phi)$.

  The spherical coordinates r_{ij}, ϕ and θ are the radial distance, azimuth and inclination, respectively. It is important to recall that the AB model uses unity radial distances between residues, that is, unit-length bond ($r_{ij} = |\hat{b}_i| = 1$).

  The initial velocities are assigned to random directions and a fixed magnitude based on the temperature. They are also adjusted to ensure that the center of mass velocity is zero. The initial accelerations are zeroed.

- **Compute forces on all particles**: this section is based on [20]. The forces f_i that act on the particles are usually derived from the potential energy. The force corresponding to $u(r)$ is $f = \nabla u(r)$. The equations of motion according to Newton's second law, $m\ddot{r}_i = \sum_{j=1(j\neq i)}^{N} f_{ij}$. Where, N represents the number of amino acids; $f_{ji} = -fij$, according to Newton's third law. Thus, each particle pair need to be examined only once. The AB model does not represent the mass value of residues. Thus, we used the unity mass in this work.

  As shown in Equation 1, the force field has three terms: bond-angle forces, bond-torsion forces and forces corresponding to the Lennard-Jones potential.

The Force that the jth residue exerts on the ith residue, corresponding to the Lennard-Jones potential is:

$$f_{ij} = 48 * \varepsilon(\sigma_i, \sigma_j)(r_{ij}^{-14} - \frac{1}{2}r_{ij}^{-8}) * r_{ij} \tag{2}$$

A change in the bond-angle (τ_i) produces forces on three neighbor residues $j = i - 2, i - 1, i$ given by:

$$- \nabla_{r_j} u(\tau_i) = - \frac{du(\tau)}{d(\cos\tau)}\Big|_{\tau=\tau_i} f_j^{(i)} \tag{3}$$

where $u(\tau_i)$ is the angle potential and $f_j^{(i)} = \nabla_{r_j} \cos(\tau_i)$
As $\sum_j f_j = 0$, the forces can be expressed by:

$$f_{i-2}^{(i)} = (c_{i-1,i-1}c_{ii})^{-1/2}[\boldsymbol{b_{i-1}}(c_{i-1,i}/c_{i-1,i-1}) - \boldsymbol{b_i}] \tag{4}$$

$$f_i^{(i)} = (c_{i-1,i-1}c_{ii})^{-1/2}[\boldsymbol{b_{i-1}} - \boldsymbol{b_i}(c_{i-1,i}/c_{ii})] \tag{5}$$

$c_{i,j}$ represents the scalar product of the ith and the jth bond vectors and it is represented by the vector $c_{i,j} = \boldsymbol{b_i} \cdot \boldsymbol{b_j}$.

The potential associated with the bond angles for the AB protein model (E_{Angles}) is shown in Equation 1. This equation can be written in cosine form because the AB model uses unit-length bonds, as follows:

$$u(\tau_i) = -k_1\hat{b}_i \cdot \hat{b}_{i+1} = -k_1 * \cos(\tau_i) \tag{6}$$

and the derivative used for the forces is given by $-\frac{du(\tau)}{d(\cos\tau)} = -k_1$.
The force associated with a torsional degree of freedom is defined in terms of the relative coordinates of four consecutive residues.

The torque caused by a rotation about the ith bond generates forces on four neighbor residues ($j = i-2, ..., i+1$) and is defined as shown in Equation 3, but replacing the argument τ_i by α_i. Where $u(\alpha_i)$ is the angle potential and $\boldsymbol{f}_j^{(i)} = \nabla_{r_j} \cos(\alpha_i)$.
As $\sum_j f_j = 0$, the forces can be expressed by:

$$\boldsymbol{f}_{i-1}^{(i)} = -(1 + c_{i-1,i}/c_{ii})\boldsymbol{f}_{i-2}^{(i)} + (c_{i,i+1}/c_{ii})\boldsymbol{f}_{i+1}^{(i)} \tag{7}$$

$$\boldsymbol{f}_i^{(i)} = (c_{i-1,i}/c_{ii})\boldsymbol{f}_{i-2} - (1 + c_{i,i+1}/c_{ii})\boldsymbol{f}_{i+1}^{(i)} \tag{8}$$

$$\boldsymbol{f}_{i-2}^{(i)} = \frac{c_{ii}}{q_i^{1/2}(c_{i-1,i-1}c_{ii} - c_{i-1,i}^2)}[t_1\boldsymbol{b_{i-1}} + t_2\boldsymbol{b_i} + t_3\boldsymbol{b_{i+1}}] \tag{9}$$

$$\boldsymbol{f}_{i+1}^{(i)} = \frac{c_{ii}}{q_i^{1/2}(c_{ii}c_{i+1,i+1} - c_{i,i+1}^2)}[t_4\boldsymbol{b_{i-1}} + t_5\boldsymbol{b_i} + t_6\boldsymbol{b_{i+1}}] \tag{10}$$

where:

$$t_1 = c_{i-1,i+1}c_{ii} - c_{i-1,i}c_{i,i+1} \tag{11}$$

$$t_2 = c_{i-1,i-1}c_{i,i+1} - c_{i-1,i}c_{i-1,i+1} \tag{12}$$

$$t_3 = c_{i-1,i}^2 - c_{i-1,i-1}c_{ii} \tag{13}$$

$$t_4 = c_{ii}c_{i+1,i+1} - c_{i,i+1}^2 \tag{14}$$

$$t_5 = c_{i-1,i+1}c_{i,i+1} - c_{i-1,i}c_{i+1,i+1} \tag{15}$$

$$t_6 = -t1 \tag{16}$$

$$q_i = (c_{i-1,i-1}c_{ii} - c_{i-1,i}^2)(c_{ii}c_{i+1,i+1} - c_{i,i+1}^2) \tag{17}$$

The potential associated with torsion for the AB protein model ($E_{torsion}$) is shown in Equation 1. This equation can also be written in cosine form as shown in Equation 6. The derivative used for the forces is given by $-\frac{du(\alpha)}{d(cos\alpha)} = -k_2$.

Further information about bond-angle and bond-torsion forces calculation (with an example of an alkane chain) can be found in [20].

– **Integrate equations of motion**: in this work, we use the velocity-verlet algorithm [23]. The implementation scheme of this algorithm is:

$$r_i(t + \delta t) = r_i(t) + v(t)\delta t + \frac{1}{2}a(t)\delta t^2 \tag{18}$$

$$v_i(t + \delta t/2) = v_i(t) + \frac{1}{2}\delta t a_i(t) \tag{19}$$

$$v_i(t + \delta t) = v_i(t + \delta t/2) + \frac{1}{2}a_i(t + \delta t)\delta t \tag{20}$$

Where, $r_i(t)$, $v_i(t)$ and $a_i(t)$ are the position, velocity and acceleration of the ith residue, respectively.

– **Perform ensemble control**: our MD simulation performs the canonical ensemble (also referred to as the ensemble NVT), where the number of particles (residues), the volume and the temperature are controlled at desired values. The temperature is controlled using the method of weak coupling to a thermal bath proposed by [6]. In this approach, coupling removes or adds energy to the system to maintain an approximately constant temperature. The velocities are scaled ($v_i(t) = \lambda * v_i(t)$) at each step using the scaling factor α:

$$\lambda = \sqrt{1 + \frac{\delta t}{\tau_T}(\frac{T_{sp}}{T} - 1)} \tag{21}$$

Where: λ, τ_T, T_{sp}, T are the scaling factor, the coupling constant, the desired temperature (set-point) and the current temperature, respectively.

– **Compute geometric constraints** a protein with the AB model is subject to geometrical constraints due to the fixed unit-length bonds between residues ($|r_i - r_j|^2 = b_{ij}^2 = 1$).

Considering a protein with N residues, there are a total of $n_c = N - 1$ geometric constraints. In this work, we use the SHAKE algorithm [21] to deal with constraints. The precision of the SHAKE algorithm is given by $|r_0 - r|/|r_0| < 10^{-k}$, where 10^{-k} is the desired precision. Our implementation has a precision of 10^{-6}.

– **Compute the desired physical quantities**: Besides the total energy of the obtained conformation, we also compute the radius of gyration [9]. Radius of gyration is a measure of compactness of a set of points (in this case, the residues of the protein). The more compact the set of points, the smaller the radius of gyration is. The radius of gyration is computed as follows

$$Rg = \sqrt{\frac{\sum_{i=1}^{N}[(x_i - \overline{X})^2 + (y_i - \overline{Y})^2 + (z_i - \overline{Z})^2]}{N}} \tag{22}$$

In this equation, x_i, y_i and z_i are the coordinates of the residues. $\overline{X}$, $\overline{Y}$ and $\overline{Z}$ are the average of all x_i, y_i and z_i; and N is the number of residues.

– General comments: our implementation uses the periodic boundary conditions [20]. We do not use real physical units because they are not defined for the AB model of proteins. Thus, the energy, temperature and length are shown in reduced units ϵ, ϵ/k_B, σ, respectively, where ϵ represents the strength of residue interactions and K_B is the Boltzmann constant.

6   Computational Experiments and Results

All experiments reported in this work were run in a desktop computer with a Intel processor Core Duo, running Linux. The application was developed in the ANSI-C programming language.

The basic MD parameters used in all the experiments are: time-step: $\delta t = 0.0001$; stop criterion: $t_{max} = 300$; and coupling constant (Berendsen's thermostat): $\tau_p = 0.01$.

Table 1 shows the synthetic sequences that were used in the experiments reported below. In this table, the second and third columns identify, respectively, the number of amino acids and the sequence of amino acids of the proteins. The first four sequences ($id = 1, ..., 4$) were used by [2,14,15] for the 3D version of the AB model. To the best of our knowledge, the other five sequences were only used for simple HP models. For instance, [4] used these sequences for the 3DHP side-chain model.

6.1   Temperature Dependence

In order to study the temperature dependence of the protein folding, experiments were done under different values of the environment temperature ($T = [0.1; 0.2; 0.5; 0.8, 1.0, 1.1, 1.2, 1.5, 1.8, 2.0]$) and using the 13-amino-acid-long sequence. The energy and radius of gyration of the best conformation obtained were recorded in order to analise the thermodynamic behavior of the protein. For each temperature, 10 independent runs were done using different unfolded proteins, and the average results are shown in Figures 2(a) and 2(b).

The overall size of the molecule (compactness of the residues), as measured by R_g, increases substantially when T increases as shown in Figure 2(b). The total

Table 1. Benchmark sequences for the AB off-lattice model

id	N	Sequence
1	13	$(ABB)^2(AB)^2BAB$
2	21	$(BA)^2B^2ABA(BBA)^2BAB^2AB$
3	34	$(ABB)^2(AB)^2(BABBA)^3B^2ABAB^2AB$
4	55	$(BA)^2B^2ABA(BBA)^2(BABBA)^2B^2ABA(BBA)^2(BABBA)^2B^2ABAB^2AB$
5	27	$AB^4A^4B^2ABABA^3BAB^2A^2B^2A$
6	27	$ABBBAAAABABAABBBABAABABBBAB$
7	27	$AB(AABB)^2A^4(BBBA)^2A^2B^2A$
8	31	$(AAB)^2A^6(BBAAAAA)^2A^2$
9	36	$BA(BBA)^{11}B$

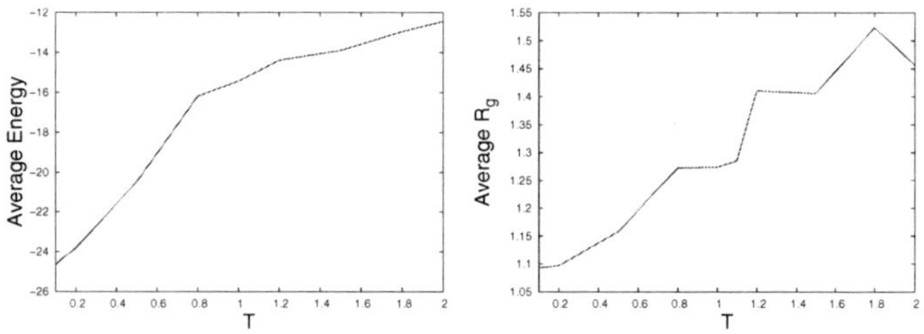

(a)   (b)

Fig. 2. Thermodynamic properties: average E (a) and R_g (b) (in dimensionless MD units)

energy also increases when T increases. This indicates a denaturation process. The best results were obtained at $T = 0.1$. Therefore, this value was fixed for the remaining experiments.

6.2   Results

Results are shown in Table 2. In this table, the third and last columns identify, respectively, the best results obtained and the average processing time. For comparison purpose, the second and fourth columns show the best results obtained for the 3D AB model of the literature and the percent difference between the best results of the literature and ours.

From the percent difference, it is observed that our results are slightly worse than the best results found in the literature. Probably, even better results were not found due to the high energy barrier between the unfolded and folded states.

Overall, the processing time is a function of the length of the sequence, growing as the number of amino acids of the sequence increases. This fact, by itself, strongly suggests the need for highly parallel approaches for dealing with the PFP.

An example of the folding pathway of the 13-amino-acid-long is presented in Figure 3. In this Figure, it is shown six folding states that were obtained in a simulation. The figure captions below each protein structure show the energy (E) and radius of gyration (R_g) at different times (t).

Table 2. Results for the 3D AB off-lattice model. N/A = not available.

N	E (Best literature)	E (best)	diff (%)	$t_p(s)$
13	-26.507 [15]	-26.4661	0.15	131.34
21	-52.917 [2]	-51.7720	2.18	234.90
34	-97.7321 [14]	-91.3662	6.73	326.50
55	-173.9803 [14]	-160.9863	7.75	504.39
27	N/A	-75.8225	–	302.56
27	N/A	-73.0161	–	264.34
27	N/A	-74.3461	–	325.72
31	N/A	-103.4963	–	247.09
36	N/A	-94.0439	–	271.53

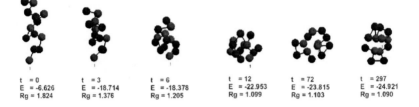

t = 0
E = -6.626
Rg = 1.824

t = 3
E = -18.714
Rg = 1.376

t = 6
E = -18.378
Rg = 1.205

t = 12
E = -22.953
Rg = 1.099

t = 72
E = -23.815
Rg = 1.103

t = 297
E = -24.921
Rg = 1.090

Fig. 3. Sample snapshot from a folding pathway.The number 1 denotes the N terminus of the chain.

The folding process starts with a denatured conformation with high energy. The protein folds through a series of intermediate states, where fragments start to pack and the protein leaves the misfolded (or partially-structured) intermediate states and forms a native-like structure. In order to reinforce our observation about the folding pathway, Figures 4(a) and 4(b) show the time dependence of the total energy and the radius of gyration.

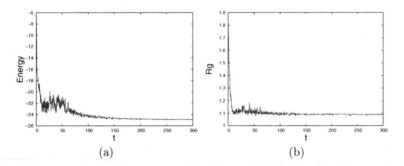

(a)

(b)

Fig. 4. $energy(t)$ (a) and $r_g(t)$(b) (in dimensionless MD units)

7   Conclusions and Future Works

The PFP is still an open problem for which there is no closed computational solution. While most works used HP models, the off-lattice AB model is still poorly explored despite being a simplified model width more biological expressiveness.

To the best of our knowledge, this paper presents the first implementation of Molecular Dynamics for the off-lattice AB model. This work also offered new reference values for five benchmark sequences that can be used in the future by other researchers.

In a broader sense, the present approach can be used to study the thermodynamics and kinetics of biophysical processes, such as the reconstruction of Transition States Ensemble (TSE) [17] and the study of sequence determinants of folding pathways [18].

Future work will include simulations and analysis of folding pathways using AB model structures built from real protein structures extracted from the PDB. Regarding the high processing time, future work will also investigate parallel versions of the MD approach. For instance, we will consider the use of a GPGPU (General Purpose Graphics Processing Units) and hardware-based accelerators [5].

Overall, results lead to interesting insights and suggest the continuity of the work. We believe that the use of Molecular Dynamics for the PFP using off-lattice AB model is very promising for this area of research.

Acknowledgments. This work is partially supported by the Brazilian National Research Council – CNPq, under grant no. 305669/2010-9 to H.S.Lopes and CAPES-DS scholarships to C.M.V. Benítez. The authors are indebted to Rafael Barreto for enlightening discussions about the basics of MD.

References

1. Anfinsen, C.B.: Principles that govern the folding of protein chains. Science 181(96), 223–230 (1973)
2. Bachmann, M., Arkm, H., Janke, W.: Multicanonical study of coarse-grained off-lattice models for folding heteropolymers. Physical Review E 71, 1–11 (2005)
3. Baker, D.: A suprising simplicity to protein folding. Nature 405, 39–42 (2000)
4. Benítez, C.M.V., Lopes, H.S.: Hierarchical parallel genetic algorithm applied to the three-dimensional HP side-chain protein folding problem. In: Proc. of the IEEE Int. Conf. on Systems, Man and Cybernetics, pp. 2669–2676 (2010)
5. Benítez, C.M.V., Scalabrin, M., Lopes, H.S., Lima, C.R.E.: Reconfigurable Hardware Computing for Accelerating Protein Folding Simulations Using the Harmony Search Algorithm and the 3D-HP-Side Chain Model. In: Xiang, Y., Cuzzocrea, A., Hobbs, M., Zhou, W. (eds.) ICA3PP 2011, Part II. LNCS, vol. 7017, pp. 363–374. Springer, Heidelberg (2011)
6. Berendsen, H.J.C., Postma, J.P.M., van Gusteren, W.F., DiNola, A., Haak, J.R.: Molecular dynamics with coupling to an external bath. Journal of Chemical Physics 81, 3684 (1984)

7. Crescenzi, P., Goldman, D., Papadimitrou, C., Piccolboni, A., Yannakakis, M.: On the complexity of protein folding. Journal of Computational Biolology 5, 423–446 (1998)
8. Dill, K.A., Bromberg, S., Yue, K., Fiebig, K.M., et al.: Principles of protein folding - a perspective from simple exact models. Protein Science 4(4), 561–602 (1995)
9. Grosberg, A.Y., Khokhlov, A.R.: Statistical Physics of Macromolecules. AIP Press (1994)
10. Gruebele, M.: Protein folding: the free energy surface. Current Opinion in Structural Biology 12(1), 161–168 (2002)
11. Hunter, L.: Artificial Intelligence and Molecular Biology, 1st edn. AAAI Press, Boston (1993)
12. Irback, A., Peterson, C., Potthast, F., Sommelius, O.: Local interactions and protein folding: A three-dimensional off-lattice approach. Journal of Chemical Physics 1, 273–282 (1997)
13. Karplus, M.: The Levinthal paradox: yesterday and today. Folding & Design 2(4), S69–S75 (1997)
14. Kim, S.Y., Lee, S.B., Lee, J.: Structure optimization by conformational space annealing in an off-lattice protein model. Physical Review E 72, 1–6 (2005)
15. Liang, F.: Annealing contour monte carlo algorithm for structure optimization in an off-lattice protein model. Chemical Physics 120, 6756–6763 (2004)
16. Liwo, A., Khalili, M., Scheraga, H.A.: Ab initio simulations of protein-folding pathways by molecular dynamics with the united-residue model of polypeptide chains. Proceedings of the National Academy of Sciences 102(7), 2362–2367 (2005)
17. Mirny, L., Shakhnovich, E.: Protein foding theory: From lattice to all-atom models. Annual Review of Biophysics and Biomolecular Structure 30, 361–396 (2001)
18. Nishimura, C., Lietzow, M.A., Dyson, H.J., Wright, P.E.: Sequence determinants of a protein folding pathway. Journal of Molecular Biology 351, 383–392 (2005)
19. Day, R., Daggett, V.: All-atom simulations of protein folding and unfolding. Advances in Protein Chemistry 66, 373–403 (2003)
20. Rapaport, D.C.: The Art of Molecular Dynamics Simulation. Cambridge University Press (2004)
21. Ryckaert, J.P., Ciccotti, G., Berendsen, H.J.C.: Numerical integration of the cartesian equations of motion of a system with constraints: Molecular dynamics of n-alkanes. Journal of Computational Physics 23, 327–341 (1977)
22. Stillinger, F.H., Head-Gordon, T.: Collective aspects of protein folding illustrated by a toy model. Physical Review E 52(3), 2872–2877 (1995)
23. Swope, W.C., Andersen, H.C., Berens, P.H., Wilson, K.R.: A computer simulation method for the calculation of equilibrium constants for the formation of physical clusters of molecules: Application to small water clusters. The Journal of Chemical Physics 76, 637 (1982)

A Bioconductor Based Workflow for Z-DNA Region Detection and Biological Inference

Halian Vilela[1], Tainá Raiol[1], Andrea Queiroz Maranhão[1],
Maria Emília Walter[2], and Marcelo M. Brígido[1]

[1] Department of Cellular Biology, Institute of Biology, University of Brasilia,
70910-900, Brasília, DF, Brazil
brigido@unb.br

[2] Department of Computer Science, Institute of Exact Sciences,
University of Brasilia, 70910-900, Brasília, DF, Brazil

Abstract. Z-DNA is an alternative conformation of the DNA molecule implied in regulation of gene expression. However, the exact role of this structure in cell metabolism is not yet fully understood. Here we present a novel Z-DNA analysis workflow using the R software environment which aims to investigate Z-DNA forming regions (ZDRs) throughout the genome. It combines thermodynamic analysis of the well-known software **Z-Catcher** with biological data manipulation capabilities of several **Bioconductor** packages. We employed our methodology in the human chromosome 14 as a case study. With that, we established a correlation of ZDRs with transcription start sites (TSSs) which is in agreement with previous reports. In addition, our workflow was able to show that ZDRs which are positioned inside genes tend to occur in intronic sequences rather than exonic and that ZDRs upstream to TSSs may have a positive correlation with the up-regulation of RNA polymerase activity.

Keywords: Z-DNA, ZDR, Z-Catcher, R, Bioconductor.

1 Introduction

The knowledge of the distribution of Z-DNA forming regions (ZDRs) throughout a genome is a valuable resource that helps to elucidate the role of this alternative DNA form in biological systems. Several evidences indicate that these regions may account for some level of gene expression regulation [6,17]. Although, it remains a challenge to determine Z-DNA genomic distribution and the regulatory networks involved in Z-DNA formation at ZDRs. Xiao *et al.* [21] developed a Z-DNA map for the human genome by searching whole chromosomes for ZDRs and locating them in relation to the transcription start sites (TSSs) of all the annotated gene models available at that moment. Although it was shown that Z-DNA has an uneven distribution along the genome, their data lack consistent biological evidence implicating Z-DNA in gene expression. Many questions remain still opened, which makes room for further investigation.

In this study, we propose a workflow for a deep investigation of the Z-DNA distribution and the possible interaction with RNA polymerase activity. The

M.C.P. de Souto and M.G. Kann (Eds.): BSB 2012, LNBI 7409, pp. 73–83, 2012.
© Springer-Verlag Berlin Heidelberg 2012

search for Z-DNA using our proposed workflow was performed along the human chromosome 14 as a case study, due to our interest in immunology research and background in antibody engineering. The immunoglobulin heavy chain locus (IgH) is located at this chromosome, which harbours genes that code for the larger polypeptide subunit of the antibody molecule [20]. In addition to a potential way to modulate the IgH genes, the understanding of how Z-DNA modulates gene expression may provide researchers with useful strategies for genetic engineering and biopharmaceutic production, therefore contributing to broaden the toolset of this research field.

1.1   The Structure of Z-DNA

Z-DNA is an alternative conformation of the canonical DNA molecule (B-DNA). It is generally formed by varying length stretches of alternating purine-pyrimidine nucleotides exposed to conditions of high supercoiling. Which are usually formed right after the passage of the RNA polymerase during transcription [16,12]. Its structure differs dramatically from the canonical form, displaying a helix which turns to the opposite direction (left) and a backbone which forms a zig-zag, hence the name "Z". This zig-zag backbone is formed due to the alternating *syn-anti* conformation of the nucleotides. The *syn* conformation, which occurs only in Z-DNA induces changes towards the phosphate bond causing the DNA backbone, formed by such bonds, to look fairly sinuous. Figure 1 summarizes the main differences among these two conformations. Due to the unique form of Z-DNA, it exhibits some notorious properties like antigenicity and binding specificity to some proteins like ADAR1[5] and the yeast protein zuotin [22].

1.2   Z-DNA Prediction Softwares: Z-Catcher and Z-Hunt

Detecting the occurrence of Z-DNA throughout the genome was first addressed by pure biophysical conformation studies [17]. After years of research some knowledge was acquired and it was possible to create prediction methods by which Z-DNA structure could be evaluated. Z-Catcher [21] and Z-Hunt [8] are computer softwares which were developed from two of such prediction methods. Both of them focus on thermodynamic analysis of an input sequence. Basically, they search for patterns of alternating purine-pyrimidine and apply free energy calculations to assess whether a given sequence is likely to change from canonical to Z conformation (B-to-Z).

Both methods try to classify potential Z forming sequences, Z-Hunt uses a comparative measurement to do so. It outputs sequences associated with a z-score, which represents the number of random base pairs that must be scanned, on average, to find a sequence with equal or better Z-forming likelihood relative to the sequence at issue [8]. On the other hand, Z-Catcher calculates the free energy based only on sequence analysis, taking into account a user defined threshold for a parameter called superhelical density, expressed by the greek letter σ (sigma). This parameter represents the free energy needed to meet the energetic requirement of a sequence to perform the B-to-Z transition.

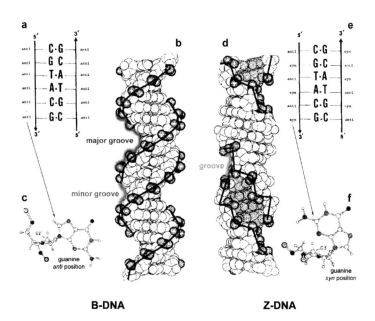

Fig. 1. Main differences between B and Z-DNA: The helix and its structural changes can be seen in **b** and **d**. *Anti* and *syn* conformations are shown respectively in **a** and **e** and their structures are detailed in **c** and **f**. [16, adapted]

Higher values of σ means that the free energy released from the relaxation of the DNA helix would easily stabilize several sequences with different Z-DNA forming potentials, while lower values tend to be more strict, meaning that the released free energy would stabilize fewer sequences. The user defines a σ value which will then be compared in each iteration with the transition required value. This process will be repeated while it stays within the threshold limits. When the limit is finally reached, the algorithm stops and the sequence is annotated as a potential ZDR.

1.3 R and Bioconductor

R is an open source software environment and programming language for statistical computing which is widely used in the scientific community. It is suited to many different tasks, such as financial, mathematical, geoprocessing and weather studies [15]. For instance, a project called Bioconductor [2], which is a repository of R add-ons packages, was created specifically for biological purposes. Most of these packages provide bioinformatic capabilities and facilitate manipulation and conversion of different formats and data.

To store and represent almost all biological sequences in this work, we used a data structure called RangedData, which is part of the Bioconductor package IRanges [13]. This structure represents biological sequences mainly by storing start-end ranges as well as information such as names, spaces (e.g. chromosomes) and other miscellaneous user-defined descriptions. The advantage of using this structure is that many other `Bioconductor` packages provide methods for comparison, trimming, flanking and sequence analysing using this format. `ChIPpeakAnno` is one example and was extensively used in this work. Its essential functionality is to find the relation between user supplied sequences and annotation data. Using its capabilities, our workflow is able to investigate the distribution of the overall distances of Z-DNA forming regions (ZDRs) relative to the nearest gene transcription start site (TSS).

All plots presented in this work were made with `ggplot2` graphics package, which is a highly customizable plotting system based on "the grammar of graphics". This system builds graphics based on separate layers overlaping each other to create a final picture, being each layer totally customizable, it is possible to create complex, nice-looking and very informative graphics.

2   Workflow Overview

Our workflow aims to provide a simple manner to analyse data allowing comparison to previous studies and bringing new information about the relationship between ZDRs and the genome. The main advantage of our approach is the direct visualization of results, eliminating the need to condense and interpret data using other softwares. Figure 2 depicts a general view of our workflow design.

The search for ZDRs is made by a slightly modified version of `Z-Catcher`, which was adapted to receive parameter input directly from the command line instead of the interactive input method of the original version. This version is integrated into R where a function may be called receiving as arguments a `fasta` file containing the sequences to be investigated and the density threshold. The function then manages all details and format converting, returning the result as an R internal object.

2.1   ZDR Distribution Analysis

Once ZDR data are available, one may build a profile of ZDRs relative to TSSs. ZDR data and an annotation file from ENCODE [18] are used as input for an R function that uses the `ChIPpeakAnno` package internally to match ZDRs and TSSs. The matches result in a distribution table of relative distances and positions which is then used to investigate where most ZDRs lie within the genome. These results will be discussed in more detail in the case study section.

2.2   ZDRs and Differential RNA Polymerase Occupation Data

In order to expand our analysis of ZDR positioning, we attempted to correlate it with ChIP-seq data regarding the RNA Polymerase II (PolII) occupation sites near the TSS. Our intent was to investigate whether ZDR sites can somehow interact with PolII and therefore change the occupation rates of the enzyme.

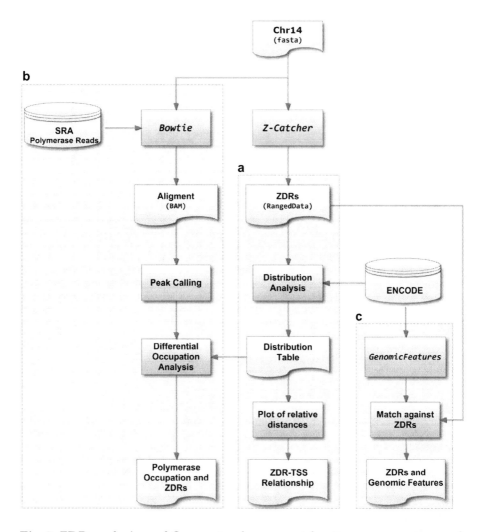

Fig. 2. ZDR analysis workflow: rectangles represent functions or processes, curved-bottom rectangles represent general data and cylinders represent database entries. Italic text represents software names or R packages and parenthesis represents file formats. Chromosome 14 `fasta` sequences are submitted to `Z-Catcher` to search for ZDRs. They are then used in the **distribution analysis (a)** by cross-matching with transcript TSS' positions from ENCODE, which results in a distribution table regarding relative distances between each ZDR and its nearest TSS. Plotting these distances gives the ZDR-TSS relationship for Chromosome 14. The same distribution table is used to search for correlations with **RNA Polymerase II differential occupation (b)** tags that were obtained by peak calling aligned reads from the SRA to Chromosome 14. At last, ENCODE transcripts are divided into **genomic features (c)** (exons, introns and splice junctions) and matched against ZDRs to investigate how they are distributed inside genes.

This analysis was carried out using public data retrieved from the NCBI Short Reads Archive (SRA) [11], particularly from Joseph *et al.* [9]. Reads derived from Illumina platform were divided in two groups: (i) cells treated with estradiol and (ii) untreated cells (control). Both groups were then aligned to the human chromosome 14 by the `Bowtie` aligner [10] with default parameters, except for `--best` which reports only the best alignments (those with fewer mismatches). The aligned reads were filtered out by a process called "peak calling" which identifies genome enriched areas where reads cluster together [14]. This process was performed in R by the `BayesPeak` package [3] and the differential occupation fold-change analysis was calculated by the `DESeq` package. [1]

2.3 ZDRs and Genomic Features

To further investigate the distribution of ZDRs in relation to genomic structure features, we searched for ZDRs within exons, introns, UTRs and intergenic regions. Although this analysis used the same `ENCODE` annotation file, it differed by being processed through the `GenomicFeatures` [4] suite so that the genes' annotation would be subdivided into their features. Then, those features were matched against the ZDRs to return an overall percentage of ZDRs relative to each genomic feature.

3 Case Study: Human Chromosome 14

3.1 Genomic Feature Analysis

As stated above, the first step of our analysis was to correlate ZDRs found by `Z-Catcher` with the gene annotation from `ENCODE`. Figure 3 shows that our approach was able to reproduce literature findings by showing an overall clustering of ZDRs around TSSs [21], as opposed to randomized distances to each TSS. Once the ZDRs are not equally spread over the genome, this distribution suggests that the ZDRs may play a role in [17] or be dependent upon transcription events [12]. To further address this issue, we looked deeper into the distribution and plotted the exact location of ZDRs relative to their nearest TSS and its respective transcript. With this analysis, we wanted to investigate if there was any bias towards specific ZDR hotspots around or within transcripts. Indeed it is possible to observe in Figure 4 that ZDRs seem to be more concentrated upstream of TSSs, which would corroborate the hypothesis on the Z-DNA relationship with transcription events.

To date, no Z-DNA mapping approach has focused on the ZDR distribution throughout the genome in relation to its genomic features. Taking that into account, it is important to further investigate this correlation, since it may reveal some unknown distribution pattern and may also help to elucidate Z-DNA function. Some works had suggested that the presence of ZDRs within introns would enable and guide the coupling of proteins from the ADAR1 family, which are responsible for mRNA editing [5]. These proteins are not only known to be

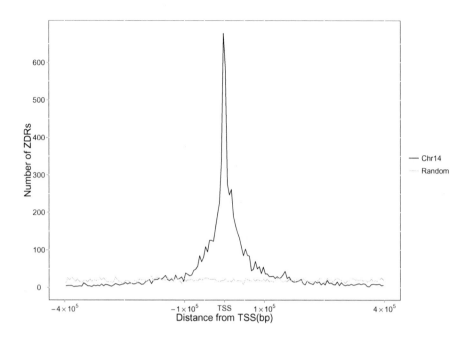

Fig. 3. Distribution of ZDRs throughout human chromosome 14: black lines show ZDR clusters around the TSS in contrast to random quasi-uniform distibution depicted in light grey

present in Z-DNA binding sites with high affinity but also to be responsible for the deamination of adenosines to inosines (which are translated as guanines). These editing events act as a source of phenotypic variation [7] and could play an important role at modulation of the nervous system [19]. If it is found that this interaction is dependent on Z-DNA formation, an important function for ZDRs would be revealed.

ZDRs found inside transcripts lie almost exclusively within introns, which account for roughly 18% of the total number of detected ZDRs (Figure 4). Considering that genes are composed mostly by intronic sequences, this percentage may not represent a strict preference of the ZDRs' distribution. Anyhow, in Table 1 it is possible to see five of the transcripts and their associated genes, which exhibit the largest number of ZDRs within introns. Such genes may be good candidates for further investigation of ADAR1 family mechanism of action, which would contribute to understand the potential role of Z-DNA guiding RNA editing enzymes.

3.2 Differential Occupation Analysis

Even though we were able to show the correlation between ZDRs and TSSs, it still remained unclear if it represents only a by-product of gene transcription

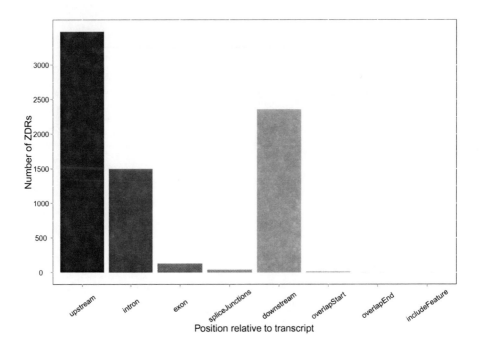

Fig. 4. Exact location of ZDRs in relation to transcripts: `includeFeature` means that a ZDR is larger enough do embrace the whole transcript, a rare situation. `intron`, `exon` and `spliceJunctions` represents ZDRs which fall within transcripts, where `spliceJunctions` represents those ZDRs shared both by introns and exons.

Table 1. Chromosome 14 transcripts with the largest number of inside-intron ZDRs

transcript	gene	# of ZDRs
ENST00000330071.6	NRXN3	151
ENST00000332068.8	NRXN3	149
ENST00000335750.5	NRXN3	125
ENST00000488612.1	RAD51L1	82
ENST00000346562.2	NPAS3	71

events or if ZDRs indeed act as gene expression regulators. To address this point, we analysed whether ZDR presence could modify RNA Polymerase II occupation of transcription start sites.

Our dataset of PolII was taken from a ChIP-Seq experiment which analysed the occupation of the promoter region of the ER-α estrogen receptor of MCF-7 cells in two conditions: activated (with presence of its ligand, estradiol) and inactivated (controlled set). With this analysis, we were able to investigate whether the differential enrichment of the PolII tags in specific locations may correlate to the presence of ZDRs.

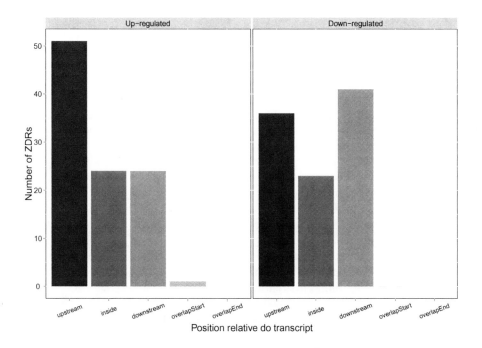

Fig. 5. Number of ZDRs overlapped by PolII tags at the 100 topmost differentially occupied transcripts: note that ZDRs tend to gather upstream of TSSs in up-regulated regions and downstream in down-regulated ones

We divided our fold-change results in two groups, one with transcripts with up-regulated occupation in the activated condition (fold-change ≥ 2) and the other with transcripts with down-regulated occupation (fold-change < 0.5). Next, we ranked the 100 topmost differentially occupied transcripts of each group and analysed their ZDRs related position. Figure 5 shows that the number of ZDRs overlapped by PolII tags at the 100 topmost differentially occupied transcripts exhibits an interesting pattern. Those ZDRs present in up-regulated genes tend to gather upstream of TSSs while those ZDRs present in down-regulated genes exhibit a weak tendency to gather downstream. This may suggest a positive correlation between ZDRs position and the activation state of genes, considering that the formation of ZDRs upstream could favour gene transcription, while ZDRs downstream could inhibit it. Nevertheless, our data need additional experimental evidence and further statistical analysis to confirm these statements.

4   Conclusions

In this work we developed a workflow for analysis of ZDR regions in animal cells that merges *in silico* data with experimental ones. The workflow uses several bioconductor packages and retrieves biological data from high throughput sequencing. Hence, one could easily correlate data from ChIP-seq and RNA-seq to ZDR regions in whole chromosomes. Our case study focused on the human chromosome 14, and the results showed that our workflow approach was able to conduct ZDR distribution analysis that corroborates previous studies. It brought as well new information on how those ZDRs spread over the chromosomal sequences.

The role of Z-DNA in gene regulation has been debated for a long time. In our case study we showed that the majority of ZDRs appear upstream of the transcripts. We also showed that when accounted for internal genomic features, ZDRs tend to concentrate in introns rather than exons. Although this was expected, it showed that our approach is able to succesfully detect ZDRs' distribution within transcripts. Hence, one could investigate in other human chromosomes or another species genome the hypothesis of ZDRs serving as anchor sites for Z-DNA binding factors such as ADAR1, which is responsible for RNA edition.

The comparison of ZDRs prediction to PolII occupancy in steroid regulated genes suggests differences in ZDRs positioning in relation to TSSs. Up regulated genes seem to concentrate ZDRs upstream of TSSs as opposed to down regulated genes that tends to concentrate ZDRs downstream. However, further experimental studies and statistical investigation are still necessary to convincingly correlate ZDRs to gene expression.

We are presently working to assemble the R scripts developed for this workflow in a user friendly R package, where the user will be able to perform similar analysis as those previously shown. Our goal is to deliver an easy and fast way to perform basic distribution analysis associated to biological information in different kinds of genomes, allowing for an efficient computing platform for the Z-DNA biology researcher.

References

1. Anders, S., Huber, W.: Differential expression analysis for sequence count data. Genome Biology 11, R106 (2010),
   http://www.bioconductor.org/packages/release/bioc/html/DESeq.html
2. Bioconductor: Open Source Software for Bioinformatics (2011),
   http://www.bioconductor.org/
3. Cairns, J., Spyrou, C., Stark, R., Smith, M.L., Lynch, A.G., Tavaré, S.: BayesPeaK–an R package for analysing ChIP-seq data. Bioinformatics 27(5), 713–714 (2011)
4. Carlson, M., Pages, H., Aboyoun, P., Falcon, S., Morgan, M., Sarkar, D., Lawrence, M.: GenomicFeatures: Tools for making and manipulating transcript centric annotations (2011), http://www.bioconductor.org/packages/release/bioc/html/GenomicFeatures.html
5. Herbert, A., Lowenhaupt, K., Spitzner, J., Rich, A.: Chicken double-stranded RNA adenosine deaminase has apparent specificity for Z-DNA. Proc. Natl. Acad. Sci. USA 92(16), 7550–7554 (1995)
6. Herbert, A., Rich, A.: The biology of left-handed Z-DNA. The Journal of Biological Chemistry 271(20), 11595–11598 (1996)
7. Herbert, A.: RNA editing, introns and evolution. Trends in Genetics 12(1), 6–9 (1996)

8. Ho, P.S., Ellison, M.J., Quigley, G.J., Rich, A.: A computer aided thermodynamic approach for predicting the formation of Z-DNA in naturally occurring sequences. The EMBO Journal 5(10), 2737–2744 (1986)
9. Joseph, R., Orlov, Y.L., Huss, M., Sun, W., Kong, S.L., Ukil, L., Pan, Y.F., Li, G., Lim, M., Thomsen, J.S., Ruan, Y., Clarke, N.D., Prabhakar, S., Cheung, E., Liu, E.T.: Integrative model of genomic factors for determining binding site selection by estrogen receptor-α. Molecular Systems Biology 6(456), 456 (2010)
10. Langmead, B., Trapnell, C., Pop, M., Salzberg, S.: Ultrafast and memory-efficient alignment of short DNA sequences to the human genome. Genome Biology 10(3), R25 (2009), `http://bowtie-bio.sourceforge.net/index.shtml`
11. Leinonen, R., Sugawara, H., Shumway, M.: The sequence read archive. Nucleic Acids Research 39(Database issue), D19–D21 (2011), `http://www.ncbi.nlm.nih.gov/sra`
12. Liu, L.F.: Supercoiling of the DNA Template during Transcription. Proceedings of the National Academy of Sciences 84(20), 7024–7027 (1987)
13. Pages, H., Aboyoun, P., Lawrence, M.: IRanges: Infrastructure for manipulating intervals on sequences, `http://www.bioconductor.org/packages/release/bioc/html/IRanges.html`, r package version 1.12.6
14. Pepke, S., Wold, B., Mortazavi, A.: Computation for chip-seq and RNA-seq studies. Nature Methods 6(11s), S22–S32 (2009)
15. R Development Core Team: R: A Language and Environment for Statistical Computing. R Foundation for Statistical Computing, Vienna, Austria (2011), `http://www.R-project.org`, ISBN 3-900051-07-0
16. Rich, A., Nordheim, A., Wang, A.H.: The chemistry and biology of left-handed Z-DNA. Ann. Rev. Biochem. 53, 791–846 (1984)
17. Rich, A., Zhang, S.: Z-DNA: the long road to biological function. Nature Reviews. Genetics 4, 566–573 (2003)
18. Rosenbloom, K.R., Dreszer, T.R., Pheasant, M., Barber, G.P., Meyer, L.R., Pohl, A., Raney, B.J., Wang, T., Hinrichs, A.S., Zweig, A.S., Fujita, P.A., Learned, K., Rhead, B., Smith, K.E., Kuhn, R.M., Karolchik, D., Haussler, D., Kent, W.J.: ENCODE whole-genome data in the UCSC Genome Browser. Nucleic Acids Research 38(Database issue), D620–D625 (2010)
19. Sommer, B., Köhler, M., Sprengel, R., Seeburg, P.H.: RNA editing in brain controls a determinant of ion flow in glutamate-gated channels. Cell 67(1), 11–19 (1991)
20. Tomlinson, I.M., Cook, G.P., Walter, G., Carter, N.P., Riethman, H., Buluwela, L., Rabbitts, T.H., Winter, G.: A complete map of the human immunoglobulin VH locus. Annals of the New York Academy of Sciences 764(1), 43–46 (1995)
21. Xiao, J., Dröge, P., Li, J.: Detecting Z-DNA Forming Regions in the Human Genome. In: International Conference on Genome Informatics 2008 (2008)
22. Zhang, S., Lockshin, C., Herbert, A., Winter, E., Rich, A.: Zuotin, a putative Z-DNA binding protein in saccharomyces cerevisiae. EMBO J. 11(10), 3787–3796 (1992)

A Probabilistic Model Checking Approach to Investigate the Palytoxin Effects on the Na$^+$/K$^+$-ATPase

Fernando A.F. Braz[1], Jader S. Cruz[2],
Alessandra C. Faria-Campos[1], and Sérgio V.A. Campos[1]

[1] Department of Computer Science, Federal University of Minas Gerais
{fbraz,alessa,scampos}@dcc.ufmg.br
[2] Biochemistry and Immunology Department, Federal University of Minas Gerais
jcruz@icb.ufmg.br
Av. Antônio Carlos, 6627, Pampulha, 30123-970 Belo Horizonte, Brazil

Abstract. Probabilistic Model Checking (PMC) is a technique that is used for the specification and analysis of unpredictable and complex systems. It can be applied directly to biological systems that show these characteristics. In this paper, PMC is used to model and analyze the effects of the palytoxin toxin (PTX) in transmembrane ionic transport systems, cellular structures responsible for exchanging ions through the plasma membrane. The correct behavior of these systems is necessary for all animal cells, otherwise the individual could present diseases and syndromes. We have discovered that high concentrations of ATP could inhibit PTX action, therefore individuals with ATP insufficiency, such as brain disorders (i.e. stroke), are more susceptible to the toxin. This type of analysis can provide a better understanding of how cell transport systems behave, give a better comprehension of these systems, and can lead to the discovery and development of new drugs.

Keywords: Probabilistic Model Checking, Systems Biology, Sodium-Potassium Pump, Palytoxin, Ion Channels Blockers and Openers.

1 Introduction

Probabilistic Model Checking (PMC) is a technique to model and analyze unpredictable and complex systems. PMC explores a stochastic model exhaustively and automatically, verifying if it satisfies properties given in special types of logics. Properties can be expressed, for example, as "What is the probability of an event happening?", offering valuable insight over the model [6,18,15].

PMC can be applied directly to biological systems to obtain a better understanding than others methods, such as simulations, which present local minima problems that PMC avoids [8,14,13,7].

We present and evaluate a PMC model of the sodium-potassium pump (or Na$^+$/K$^+$-ATPase), a transmembrane ionic transport system that exists in all

M.C.P. de Souto and M.G. Kann (Eds.): BSB 2012, LNBI 7409, pp. 84–96, 2012.

animal cells and important to several biological processes i.e. cell volume control and heart muscle contraction. Its irregular behavior is related to numerous diseases and syndromes, and it is one of the main targets of toxins and drugs [3].

In our model we have exposed the pump to a deadly toxin called palytoxin (PTX) that essentially disrupts the behavior of the pump. This was done to better understand the effect of PTX interactions with the pump [19].

We have found that high doses of Adenosine Triphosphate (ATP), the cellular energy unit, could inhibit the action of PTX. For example, when the concentration of ATP is changed from 10 mM to 100 mM, the probability of being in PTX related sub-states is reduced by 38.37%. This suggests that individuals with ATP insufficiency are more susceptible to the toxin. This ATP deficiency appears in different forms, such as brain disorders, for example, stroke. Since ATP production cannot be stimulated directly, the study of its role and capability to change our Na^+/K^+-ATPase model behavior is even more important.

PMC can improve our understanding of cell transport systems and its behavior, and can lead to the discovery and development of new drugs.

Outline. This paper describes transmembrane ionic transport systems in Section 2. The related work to the analysis of these systems and PMC usage are discussed in Section 3. Our model is introduced in Section 4 and 5, and our experiments, properties and results are shown in Section 6. Finally, our conclusions and future works are presented in Section 7.

2   Preliminaries

2.1   Transmembrane Ionic Transport Systems

Transmembrane ionic transport systems are structures present in all animal cells. They are responsible for ion exchange from the extra to the intracellular medium. The difference in charges and concentrations between ions in these sides creates an electrochemical gradient. Ionic transport systems are responsible for the gradient maintenance, which is necessary for cells to perform their functions [2].

There are two types of transmembrane transport systems: ion channels, a passive transport that does not consume energy; and ionic pumps, an active transport that consumes energy in the form of Adenosine Triphosphate (ATP, or $[ATP]^i$ for intracellular ATP concentration). Ion channels depend on the concentration gradient of ions to be transported, moving in favor of that gradient. Ionic pumps move ions against the concentration gradient, electrical charge or both [16]. Once open, ion channels diffuse ions rapidly, allowing abrupt changes in ions concentrations. Ionic pumps, on the other hand, move ions very slowly, permitting only subtle changes in ions concentrations.

A generic ion channel is shown in Figure 1. The ion channel is initially closed, and there is a high concentration of ions in the extracellular medium. A signaling molecule binds to the ion channel. This opens the ion channel and allows the ions to diffuse rapidly from the extra to the intracellular side (low concentration of ions). The change in the ion concentration in the intracellular medium triggers

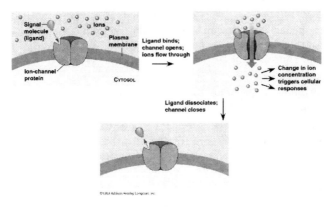

Fig. 1. An ion channel transferring ions. Adapted from [4].

a cellular response, the signaling molecule unbinds the ion channel, and the ion channel closes, interrupting ion flux.

Ion channels and ionic pumps permit the passage of only specific ions such as sodium, potassium and calcium. For ionic pumps, the passage of ions can be viewed as two gates, one internal and another external, that open or close based on different factors, such as chemical and electrical signals [2].

One ionic pump is the sodium-potassium pump (or Na^+/K^+-ATPase), shown in Figure 2. It is responsible for exchanging three sodium ions from the intracellular medium for two potassium ions from the extracellular medium. This pump can be in two conformational states: open to its inner side, and open to its outer side. Three sodium ions can bind to the pump when its opened to the intracellular side. After that, an ATP binds to the pump, which is followed by its hydrolysis, releasing the sodium ions outside. A phosphate remains bound to the pump and a molecule of Adenosine Diphosphate (ADP) is released. Two potassium ions in the outside bind to the pump, which are released in the intracellular side, as well as the phosphate. The pump now can repeat the process [2].

These cellular structures are involved in several biological processes, i.e. cellular volume control, nerve impulse, coordination of heart muscle contraction and release of accumulated calcium in the sarcoplasmic reticulum for performance of muscle contraction [2].

Ion transport systems are one of the main targets in research for discovery and development of drugs, since its irregular behavior is associated with several diseases, for example hypertension and Parkinson's disease. Cardiac glycosides are one type of these drugs, for example digoxin (also known as digitalis) and ouabain, inotropic drugs that are used to improve heart performance by increasing its contraction force [2]. Because of their important role in the nervous system, ion transport systems are also the main targets of neurotoxins [2]. One of the toxins that affect these structures is the palytoxin (PTX, or $[PTX]^o$ for extracellular PTX concentration), a deadly toxin found in corals of the *Palythoa toxica* species. The PTX affects the Na^+/K^+-ATPase, changing its behavior to the one of an ion channel [3].

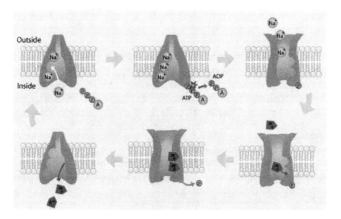

Fig. 2. The Na$^+$/K$^+$-ATPase cycle. Adapted from [11].

These structures are not well understood, despite their discovery over 50 years ago [2]. Due to their different functions and transfer rates, ion channels and ionic pumps have been seen as different entities. However, recent discoveries such as the interaction between PTX and the Na$^+$/K$^+$-ATPase, are forcing new studies about how these structures really work and how they are seen by the scientific community.

Ion channels and ionic pumps usually are investigated using experimental results in laboratory benches, which are expensive both financial and timewise. To avoid these costs, different types of simulations, mathematical and computational methods are employed, among these include ordinary differential equations (ODE) and Gillespie's algorithm for stochastic simulations [9]. Despite their ability to obtain valuable information, simulations do not cover every possible situation, and might never search certain regions of the model state space, therefore possible missing some rare and noisy events.

3   Related Work

3.1   Experimental and Simulational Techniques

The authors of [3] investigated PTX and its interactions with the Na$^+$/K$^+$-ATPase. They found that PTX drastically modifies the nature of the pump after binding to it, which changes the behavior of the pump to the one of an ion channel. They suggest that PTX could be an useful tool in experiments to discover the control mechanisms for opening and closing the gates of ion pumps. This is later visited by the authors of [19] through mathematical simulations using non-linear ODEs and considering only states and reactions related to the phosphorylation process (phosphate binding and unbinding to the pump).

Interactions of PTX with the complete model for the Na$^+$/K$^+$-ATPase are analyzed in [20]. This series of works of Rodrigues et al. can be viewed as simulational approach of the experimental results of Artigas et al. in [3]. One could view this paper as a model checking approach to the same experiments.

3.2 Model Checking

The tools used in the formal verification of biological systems that are more closely related to this work are PRISM [15], BioLab [7] and Ymer [23].

PRISM supports different types of models, properties and simulators [15]. It has been largely used in several fields, i.e. communication and media protocols, security and power management systems. We have used PRISM in this work for several reasons, which include: exact PMC in order to obtain accurate results; Continuous-time Markov Chain (CTMC) models, suited for our field of study; rich modeling language that allowed us to build our model; and finally property specification using Continuous Stochastic Logic (CSL), which is able to express qualitative and quantitative properties.

The authors of [7] introduce a new algorithm called BioLab. Instead of building explicitly all states of a model, the tool generates the minimum number of necessary simulations, given error bounds parameterized for acceptance of false positives and false negatives of the properties to be verified. This algorithm is based on the works of [23], author of the approximate model checker Ymer. We did not use BioLab or Ymer because our initial analysis demanded exact results. Only after these preliminary results we could have used an approximate analysis.

The authors illustrate in [14] the application of PMC to model and analyze different complex biological systems for example the signaling pathway of Fibroblast Growth Factor (FGF), a family of growth factors involved in healing and embryonic development. The analysis of other signaling pathways such as MAPK and Delta/Notch can be seen in [13].

The use of PMC is demonstrated also in [12], where the authors examine and obtain a better understanding of mitogen-activated kinase cascades (MAPK cascades) dynamics, biological systems that respond to several extracellular stimuli, i.e. osmotic stress and heat shock, and regulate many cellular activities, such as mitosis and genetic expression.

4   The Na^+/K^+-ATPase Model

The model is written in the PRISM language (used by the PRISM model checker) and consists of modules for each of the molecules (ATP, Phosphate or P_i and Adenosine Diphosphate or ADP) and one main module for the pump. This model does not include PTX. A small part of the model is shown in Figure 4, and its complete version can be seen in [1].

Each molecule module contains a variable to store the current number of molecules, i.e. `atpIn` for ATP. Also, there are transitions that represent reactions, which are responsible for changing the number of molecules. The list of reactions can be found in [19], and in the comments of our model [1].

Na$^+$/K$^+$-ATPase PRISM Model

```
module atp
  atpIn : [0..(N)] init ATPI; // number of ATP inside cell
  // reaction1: ATPi + E1 <-> ATPhighE1
  [r1] atpIn>=1 -> atpIn : (atpIn'=atpIn-1);
  [rr1] atpIn<=(N-1) -> 1 : (atpIn'=atpIn+1);
endmodule

module pump
  E1 : [0..1] init 1;         // e1 conformational state
  ATPhighE1 : [0..1] init 0; // e1 with atp bound to its high affinity site
  //reaction1: ATPi + E1 <-> ATPhighE1
  [r1] E1=1 & ATPhighE1=0 -> 1 : (E1'=0) & (ATPhighE1'=1);
  [rr1] E1=0 & ATPhighE1=1 -> 1 : (E1'=1) & (ATPhighE1'=0);
endmodule

// base rates
const double r1rate = 1.50*pow(10,4)/(0.001*V*AV);
const double rr1rate = 1.64;

// module representing the base rates of reactions
module base_rates
  [r1] true -> r1rate : true;
  [rr1] true ->rr1rate : true;
endmodule
```

Fig. 3. Na$^+$/K$^+$-ATPase PRISM Model

The main module controls the pump, keeping track of its current sub-state. The sub-states are a boolean vector, where only one position can and must be true. The main module also has transitions which change the pump sub-state.

The Albers-Post model [17] represents the Na$^+$/K$^+$-ATPase cycle and it can be seen on the left side of Figure 4. According to it, the pump can be in different sub-states, which change depending on different reactions involving ATP, ADP and P$_i$. The right side is the Palytoxin model, which is covered later.

The pump can be open or closed to the extra and intracellular sides. An ATP can bind to the pump in either its high or low affinity binding site. An ATP bound to the pump can be hydrolyzed, leaving one P$_i$ bound to the pump and releasing one ADP. The reactions are bidirectional, and their rates can be seen in [19].

The pump can be in two major states: open to the intracellular side (**E1**, in our PRISM model) and open to the extracellular side (**E2**). Those states are divided in different sub-states: open to the intracellular side, with an ATP bound to its high affinity binding site (**ATPhighE1**); phosphorylated and open to the intracellular side, with an ATP bound to its low affinity binding site (**ATPlowPE1**); phosphorylated and open to the extracellular side (**PE2**); open to the extracellular side, with an ATP bound to its low affinity binding site (**ATPlowE2**); and phosphorylated and open to the extracellular side, with an ATP bound to its low affinity binding site (**ATPlowPE2**).

4.1   Discrete Chemistry

The main components of our model are molecules (ATP, P_i and ADP) and the Na^+/K^+-ATPase, which can interact with each other through several elementary reactions. There is one additional molecule (PTX) in the palytoxin extension for this model, covered in the next section.

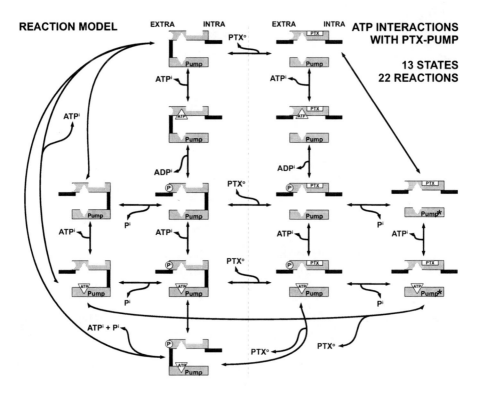

Fig. 4. The classical Albers-Post model [17]. Adapted from [19].

The concentration of each of these components is a discrete variable, instead of a continuous function. Therefore, we have converted the amount of initial concentration of molecules from molarity (M) to number of molecules. The stochastic rates for forward and backward transitions are from [19]. The substrates concentrations ($[ATP]^i = 0.00500$, $[P]^i = 0.00495$ and $[ADP]^i = 0.00006$) are from [5]. The *cell volume* is from [10].

In order to convert the initial amount of molecules given in molarity ($[X]$) into quantities of molecules ($\#X$), we have used the following biological definition:

$$\#X = [X] \times V \times N_A \tag{1}$$

where V is the cell volume and N_A is the Avogadro constant.

5   The Palytoxin Model

The palytoxin model is an extension of the Na$^+$/K$^+$-ATPase model, described in the previous section and partially shown in Figure 4. It corresponds to the right side of the Figure 4, and it is based on the description by [19] and [3].

```
Palytoxin PRISM Model

module ptx
  // number of PTX outside the cell
  ptxOut : [0..(PTXO)] init PTXO;
  //reaction p1: PTXo + E1 <-> PTXE
  [rp1] ptxOut>=1 -> ptxOut : (ptxOut'=ptxOut-1);
  [rrp1] ptxOut<=(PTXO-1) -> 1 : (ptxOut'=ptxOut+1);
endmodule

module pump
  PTXE : [0..1] init 0;
  //reaction p1: PTXo + E1 <-> PTXE
  [rp1] E1=1 & PTXE=0 -> 1 : (E1'=0) & (PTXE'=1);
  [rrp1] E1=0 & PTXE=1 -> 1 : (E1'=1) & (PTXE'=0);
endmodule

// base rates
const double rp1rate=2.73*pow(10,1)/(0.001*V*AV);
const double rrp1rate=6.0*0.0001;

// module representing the base rates of reactions
module base_rates
  [rp1] true -> rp1rate : true;
  [rrp1] true -> rrp1rate : true;
endmodule
```

Fig. 5. Palytoxin PRISM Model

This expanded model consists of one additional molecule module (PTX), additional reactions in each of the already present modules and additional sub-states for the pump module. Initial concentrations for [PTX]^o and stochastic rates for transitions between states were obtained in [19]. Once again, a fragment of the model is shown in Figure 5, and its complete version can be seen in [1].

The sub-states correspond to the pump bound to PTX, when the pump is open to both sides behaving like an ion channel. There are six additional sub-states for the pump: bound to a PTX (**PTXE**, in our model); bound to a PTX, with an ATP bound as well to its high affinity binding site (**PTXATPhighE**); bound to a PTX and the pump is phosphorylated (**PTXPE**); bound to a PTX, after the pump dephosphorylated (**PTXE***); bound to a PTX, with an ATP bound as well to its low affinity binding site, and phosphorylated (**PTXATPlowPE**); and bound to a PTX, with an ATP bound as well to its low affinity binding site, after dephosphorylation (**PTXATPlowE***).

6   Experimental Results

6.1   Parameters and Model Complexity

We can explore our model in three dimensions: $[PTX]^o$ (extracellular PTX concentration), $[ATP]^i$ (intracellular ATP concentration) and pump volume. Each dimension represents one aspect of the model, and can be changed to modify its behavior. The values of these parameters influence directly to the complexity of the model (number of states, transitions and topology), and to the time to build and verify model properties. Table 1 shows how these values increase in function of pump volume, for $[PTX]^o$ =0.001 μm, $[ATP]^i$ =10 mm and a state reward property discussed further below. The machine used to perform experiments is a Intel(R) Xeon(R) CPU X3323, 2.50GHz and has 17 GB of RAM memory.

Table 1. Model complexity, build and check time in function of pump volume

Pump Volume	States	Transitions	Build Time	Check Time
10^{-22}	376	1912	0.044 s	310.895 s
10^{-21}	1274	7140	0.081 s	321.506 s

The cellular volume in an animal cell is 10^{-12} L [10], which is prohibitive to represent since it would cause the classical problem of state space explosion for model checking. Our analysis is restricted to only one pump of the cell. As a consequence, it would also not be realistic to model a large volume because in the real cell this large volume is shared between several pumps and other cellular structures, not limited to pumps. Our abstraction reduces the cell volume concentrating our analysis in one or few pumps and their surroundings. We achieve this by maintaining the proportions between all interacting components. Therefore, our dimension for cellular volume is called pump volume and is usually 10^{-22} L. Even though those values are many orders of magnitude smaller than the real values, they still represent proper cell behavior, and can be interpreted as using a magnifying glass to investigate a portion of the cell membrane.

On the other hand, for some dimensions we have used more values than intuition suggests, ranging from three orders of magnitude below and above their literature reference values, i.e. $[PTX]^o$ and $[ATP]^i$, respectively 5 μm for $[PTX]^o$ and between 1 and 10 mM for $[ATP]^i$. This is particularly interesting because we can model different situations for pump behavior, including abnormal concentrations levels for $[ATP]^i$ due to some disease or syndrome, and different degrees of exposure to $[PTX]^o$, from mild to fatal exposure.

We formulated many properties that can be seen in [1]. Due to space limitations we have chosen to present the most important ones: state and rate rewards.

6.2   PTX Inhibition by High Doses of ATP

In order to observe what is the probability of being in PTX and non-PTX related states over time, all states and rates were labeled and quantified using rewards.

The following excerpt of the model in Figure 6.2 shows the rewards for the sub-states *PE2* and *PTXE*, respectively the pump open the extracellular side with a phosphate bound to it, and the pump open to both sides and bound to PTX. Basically rewards are incremented each time its conditions are true.

State Rewards PRISM Model
| ```
rewards "ptxe"
  (PTXE=1) : 1;
endrewards
``` |

Accumulated State Reward Property
$\mathbf{R}\{\text{"ptxe"}\}=?\ [\ \mathbf{C}{<}{=}T\ ]$ What is the expected accumulated reward for the state `ptxe` until time T?

Fig. 6. State Rewards and Accumulated State Reward Property

Since the model now has rewards for each state, we are able to count the expected quantity of the accumulated reward associated with each sub-states over time, with properties such as the one shown in Figure 6.2. Using the operator **R** we are able to quantify the reward for some given event, for example the number of times the model was in sub-state *PE2*. The operator **C** allows to quantify accumulated rewards for a given time T, therefore we are able to observe rewards over time.

Considering a single pump, a cellular volume of 10^{-22} M, $[ATP]^i = 10$ mM and $[PTX]^o = 10$ μm, at instant T=100, the expected rewards associated with the same two sub-states PE2 and PTXE are respectively 33.3543 and 1.0100. In other words, in 100 seconds, the pump is expected to be open to the extracellular side and phosphorilated approximately 34.20% of the time, and the pump is expected to be bound exclusively to PTX only 1.04% of the time.

Using a broad spectrum of different $[ATP]^i$ and $[PTX]^o$, we have found that for the cellular volume of 10^{-22} L there are only two sets of values for sub-state rewards. One set is associated with $[ATP]^i$ equals or below to 10 mM, while the other set is associated with $[ATP]^i$ above 10 mM. For example, when $[ATP]^i = 100$ mM, the expected rewards associated with the two sub-states PE2 and PTXE changes to respectively 37.3577 and 0.5903, or 39.24% and 0.62% of the time. Therefore, as we increased $[ATP]^i$, the likelihood of the pump being open to the extracellular side and phosphorilated increased 14.73%, and for the pump to be bound exclusively to PTX decreased 40.38%.

Summing the rewards associated with PTX related sub-states and the unrelated sub-states, and dividing each by the total, for $[ATP]^i = 10$ mM we found that at instant T=100, PTX related states corresponded to 24.21%. As we increased $[ATP]^i$ to 100 mM, PTX related states corresponded to only 14.92%, suffering a 38.37% reduction. Therefore, we have found that as $[ATP]^i$ increases, the probability of being in PTX related sub-states decreases. This result suggests that ATP is an inhibitor of PTX, and as consequence people with ATP insufficiency would be more vulnerable to this toxin. ATP deficiency appears
```

in different forms, i.e. brain disorders, for example, stroke and metabolic encephalopathies [22].

Similar reward structures to the ones of Figure 6.2 were created for reaction rates. For $[ATP]^i = 10$ mM we found that during the first 100 seconds PTX related reactions correspond only to approximately 8.01%. Once we change $[ATP]^i$ to 100 mM the role of PTX related reactions decreases by approximately 42.97%, which reinforces our discovery that high doses of ATP inhibit PTX action. The most active reactions are dephosphorylation, changes in the pump conformational state, and coupling and releasing of ATP. Using other cell volumes, we have found that new sets of values emerge, eventually reaching one set for each $[ATP]^i$, even though the same behavior remains.

The experimental conditions used to study the major effects of various ligands including ATP on PTX-modified Na^+/K^+-ATPase [3] are rather different and this poses a problem in terms of comparison with our results. The inhibitory effect elicited by ATP as predicted by our model has been not verified experimentally and it was totally unexpected but the result raises an important point that may be worth to explore experimentally.

Our results suggest that in the presence of palytoxin, the extent of phosphorylation from ATP is greatly reduced probably by a PTX-promoted rapid dephosphorylation step that could, at high concentrations of ATP, lead to inhibition of ATP binding. This finding is interesting because it reinforces the notion that the phosphorylated intermediates formed from ATP are different and this may change PTX affinity and the overall behavior of the Na^+/K^+-ATPase. There are some reports in the literature that could support this result [21].

7   Conclusions and Future Work

The Na^+/K^+-ATPase is a cellular structure responsible for exchanging ions through the plasma membrane. Its correct behavior is necessary for all animal cells, otherwise the health of the individual could be in risk due to diseases and syndromes. A model for a single Na^+/K^+-ATPase interacting with the toxin palytoxin (PTX) has been built using the Probabilistic Model Checking tool PRISM. PTX essentially disrupts the Na^+/K^+-ATPase regular behavior. PMC has allowed us to investigate the model, which show unpredictable and complex characteristics. Properties about biological events were expressed in probabilistic logics, e.g. "What is the probability of being in PTX related sub-states?".

We have discovered that as $[ATP]^i$ increases, the probability of being in PTX related sub-states decreases. For example, when $[ATP]^i$ is changed from 10 mM to 100 mM, the probability of being in PTX related sub-states is reduced by 38.37%. This suggests that high concentrations of $[ATP]^i$ could inhibit $[PTX]^o$ action, which implies that individuals with $[ATP]^i$ insufficiency are more susceptible to $[PTX]^o$ effects. This $[ATP]^i$ deficiency appears in different forms, such as brain disorders, for example, stroke. The study of the role and ability of $[ATP]^i$ to change our Na^+/K^+-ATPase model behavior is even more important, because the production of $[ATP]^i$ cannot be stimulated directly.

We have shown in this work that PMC can be used to obtain valuable insight of transmembrane ionic transport systems in a simple and complete way. This type of analysis can provide a better understanding of how cell transport systems behave, give a better comprehension of these systems, and can lead to the discovery and development of drugs.

Future works: confront the results with wet lab experiments; expand the model to other Albers-Post sub-states (e.g. related to potassium and sodium); explore other dimensions such as the number of pumps; and adapt our model to other toxins (for example, ouabain) or even drugs (e.g. digitalis).

References

1. http://www.dcc.ufmg.br/~fbraz/bsb2012/
2. Aidley, D.J., Stanfield, P.R.: Ion channels: molecules in action. Cambridge University Press (1996)
3. Artigas, P., Gadsby, D.C.: Large diameter of palytoxin-induced na/k pump channels and modulation of palytoxin interaction by na/k pump ligands. J. Gen. Physiol. (2004)
4. Campbell, N.A., Reece, J.B., Mitchell, L.G.: Biology, 5th edn. (1999)
5. Chapman, J.B., Johnson, E.A., Kootsey, J.M.: Electrical and biochemical properties of an enzyme model of the sodium pump. Membrane Biology (1983)
6. Clarke, E.M., Emerson, E.A.: Design and Synthesis of Synchronization Skeletons Using Branching-Time Temporal Logic. In: Kozen, D. (ed.) Logic of Programs 1981. LNCS, vol. 131, pp. 52–71. Springer, Heidelberg (1982)
7. Clarke, E.M., Faeder, J.R., Langmead, C.J., Harris, L.A., Jha, S.K., Legay, A.: Statistical Model Checking in *BioLab*: Applications to the Automated Analysis of T-Cell Receptor Signaling Pathway. In: Heiner, M., Uhrmacher, A.M. (eds.) CMSB 2008. LNCS (LNBI), vol. 5307, pp. 231–250. Springer, Heidelberg (2008)
8. Crepalde, M., Faria-Campos, A., Campos, S.: Modeling and analysis of cell membrane systems with probabilistic model checking. BMC Genomics 12(suppl. 4), S14 (2011), http://www.biomedcentral.com/1471-2164/12/S4/S14
9. Gillespie, D.T.: Exact stochastic simulation of coupled chemical reactions. The Journal of Physical Chemistry 81(25), 2340–2361 (1977)
10. Hernández, J.A., Chifflet, S.: Eletrogenic properties of the sodium pump in a dynamic model of membrane transport. Membrane Biology 176, 41–52 (2000)
11. Karp, G.: Cell and Molecular Biology, 5th edn. (2008)
12. Kwiatkowska, M., Heath, J.: Biological pathways as communicating computer systems. Journal of Cell Science 122(16), 2793–2800 (2009)
13. Kwiatkowska, M., Norman, G., Parker, D.: Quantitative Verification Techniques for Biological Processes. In: Algorithmic Bioprocesses. Springer (2009)
14. Kwiatkowska, M., Norman, G., Parker, D.: Probabilistic Model Checking for Systems Biology. In: Symbolic Systems Biology, pp. 31–59. Jones and Bartlett (2010)
15. Kwiatkowska, M., Norman, G., Parker, D.: PRISM 4.0: Verification of Probabilistic Real-Time Systems. In: Gopalakrishnan, G., Qadeer, S. (eds.) CAV 2011. LNCS, vol. 6806, pp. 585–591. Springer, Heidelberg (2011)
16. Nelson, D.L., Cox, M.M.: Lehninger Principles of Biochemistry, 3rd edn. (2000)
17. Post, R., Refyvary, C., Kume, S.: Activation by adenosine triphosphate in the phosphorylation kinetics of sodium and potassium ion transport adenosine triphosphatase. J. Biol. Chem.

18. Queille, J.P., Sifakis, J.: A temporal logic to deal with fairness in transition systems. In: 23rd Annual Symposium on Foundations of Science, SFCS'08, pp. 217–225 (1982)
19. Rodrigues, A.M., Almeida, A.C.G., Infantosi, A.F., Teixeira, H.Z., Duarte, M.A.: Model and simulation of na+/k+ pump phosphorylation in the presence of paly-toxin. Computational Biology and Chemistry 32(1), 5–16 (2008)
20. Rodrigues, A.M., Infantosi, A.F.C., Almeida, A.C.G.: Palytoxin and the sodium/potassium pump-phosphorylation and potassium interaction. Physical Biology (2009)
21. Tosteson, M., Thomas, J., Arnadottir, J., Tosteson, D.: Effects of palytoxin on cation occlusion and phosphorylation of the (na^+/k^+)-atpase. Journal of Membrane Biology 192, 181–189 (2003), 10.1007/s00232-002-1074-9
22. Yamada, K., Inagaki, N.: Atp-sensitive k+ channels in the brain: Sensors of hypoxic conditions. Physiology 17(3), 127–130 (2002)
23. Younes, H.L.S.: Ymer: A Statistical Model Checker. In: Etessami, K., Rajamani, S.K. (eds.) CAV 2005. LNCS, vol. 3576, pp. 429–433. Springer, Heidelberg (2005)

RFMirTarget: A Random Forest Classifier for Human miRNA Target Gene Prediction

Mariana R. Mendoza[1], Guilherme C. da Fonseca[2], Guilherme L. de Morais[2],
Ronnie Alves[3], Ana L.C. Bazzan[1], and Rogerio Margis[2]

[1] PPGC, UFRGS, P.O. Box 15064, Porto Alegre, RS, Brazil
{mrmendoza,bazzan}@inf.ufrgs.br
[2] PPGBCM, UFRGS, P.O. Box 15005, Porto Alegre, RS, Brazil
{guicf13,guilherme.loss}@gmail.com, rogerio.margis@ufrgs.br
[3] Vale Technological Institute Sustainable Development, Belém, PA, Brazil
ronnie.alves@vale.com

Abstract. MicroRNAs (miRNAs) are key regulators of eukaryotic gene expression whose fundamental role has been already identified in many cell pathways. The correct identification of miRNAs targets is a major challenge in bioinformatics. So far, machine learning-based methods for miRNA-target prediction have shown the best results in terms of specificity and sensitivity. However, despite its well-known efficiency in other classifying tasks, the random forest algorithm has not been employed in this problem. Therefore, in this work we present RFMirTarget, an efficient random forest miRNA-target prediction system. Our tool analyzes the alignment between a candidate miRNA-target pair and extracts a set of structural, thermodynamics, alignment and position-based features. Experiments have shown that RFMirTarget achieves a Matthew's correlation coefficient nearly 48% greater than the performance reported for the MultiMiTar, which was trained upon the same data set. In addition, tests performed with RFMirTarget reinforce the importance of the seed region for target prediction accuracy.

Keywords: miRNA, target prediction, random forest, gene regulation.

1 Introduction

MicroRNAs (miRNAs) are non-coding RNAs of ~22 nucleotides in length that act as negative regulators of gene expression, thus playing an important role in gene regulation by targeting mRNAs with cleavage or translational repression [1]. The miRNA biogenesis is similar in both animals and plants. Mature miRNAs are formed from longer primary transcripts by two sequential processing steps mediated by a nuclear and a cytoplasmic RNase III endonuclease. In animals the responsible enzymes are Drosha and Dicer, respectively, while in plants both cleavages are performed by a Dicer homolog, DCL [1]. These cleavages generate a $60-70$ nt stem-loop miRNA precursor (pre-miRNAs) and a mature miRNA duplex, respectively. Further, the mature miRNA duplex is assembled into an

M.C.P. de Souto and M.G. Kann (Eds.): BSB 2012, LNBI 7409, pp. 97–108, 2012.
© Springer-Verlag Berlin Heidelberg 2012

effector complex known as RNA-induced silencing complex (RISC). The gene repression, caused by miRNA silence, can occur in two different ways: i) by degradation of messenger RNA (mRNA) through the RNA interference (RNAi) pathway or ii) by inhibiting protein translation [2].

Since the discovery of miRNAs in *Caenorhabditis elegans* [3], in which they were related to control developmental timing, miRNAs have been characterized in both animals and plants in several metabolic processes, such as growth, apoptosis, cell proliferation, stress responses and defense against viruses and other diseases [4,5]. In humans, miRNAs play a critical role in tumorigenesis, acting either as tumor suppressors or oncogenes. Moreover, recent studies have shown that miRNAs are highly related with cancer progression, including initiating, growth, apoptosis, invasion and metastasis [6].

Two main challenges are involved in the study of miRNAs: the identification of novel miRNAs and the prediction of miRNAs targets. The existence of a stem-loop is the key feature adopted by *ab initio* prediction methods to identify novel miRNAs. In this sense, machine learning (ML) algorithms, among which support vector machine (SVM) [7,8], random forest [9] and naïve Bayes [10] stand out, have been extensively applied to the task of learning how to distinguish real pre-miRNAs from pseudo pre-miRNAs based on a set of descriptive features. Following this direction, ML methods can help in the computational prediction of miRNAs targets, although this is considered a more difficult problem.

Basically, the interaction of a miRNA and its target occurs by complementarity of their nucleotide sequences, with some functional differences between plants and animals miRNAs. Plant miRNAs bind their targets with perfect or near perfect complementarity and mostly in their open read frames (ORFs) [11]. In contrast, in most cases, animals miRNAs sequences have a partial complementarity to their targets and the hybridization occurs predominantly in the 3' untranslated regions (3' UTRs) [12]. Furthermore, in animals, a region of miRNA called *seed* plays an important role in the correct interaction between the miRNA and its target. This region comprehends six to eight nucleotides in the 5' end of the miRNA that have (almost) strict pairing with the mRNA target. The classification of miRNA target sites can be thus distinguished into three types: i) 5' dominant canonical, where a perfect match occurs in the seed region and an extensive base pairing is observed in the 3' end of the miRNA, ii) 5' dominant seed only, which presents an extensive base pairing only in the seed region and iii) 3' compensatory, in which the seed region does not contains a perfect base pairing, but the 3' end of the miRNA displays a more extensive base pairing [13,14].

Several different computational tools have been already developed for target prediction analysis [15]. In general, these tools are based on features derived from the interaction between a miRNA and its potential targets, such as seed complementarity, thermodynamics stability, presence of multiple sites and evolutionary conservation among species [14]. Among these, ML-based algorithms have had the best results so far in terms of specificity and sensitivity [16]. One well-known classifier is the random forest algorithm [17], further explained in Section 2.1.

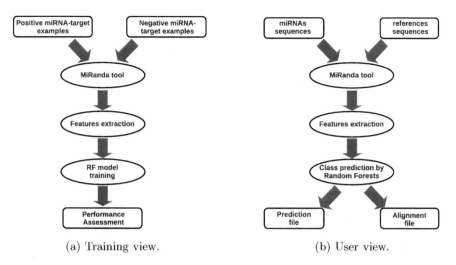

(a) Training view.   (b) User view.

Fig. 1. Fluxogram of RFMirTarget training and functioning. (a) RFMirTarget is trained with positive and negative miRNA-target examples. This data set is analized by miRanda, whose output is processed for features extraction. A RF model is then built upon these features. (b) Once trained, RFMirTarget may be used to classify unknown miRNA-targets: given two files with miRNAs and references sequences, RFMirTarget outputs the confidence and alignment for all predicted miRNA-targets.

Its popularity is due mainly to its meaningful efficiency when compared to other classification methods. This efficiency, in turn, comes from the manner the algorithm profits from ensemble predictions. Although already known as an efficient approach for identifying novel miRNAs [9], random forests have not been explored in the context of miRNA-target prediction yet. Therefore, in the present paper, we introduce the RFMirTarget tool, a random forest model for the prediction of miRNAs targets. Tests with biologically validated examples have shown that the proposed model is indeed robust and has performance superior than the state-of-the-art tool MultiMiTar [16].

2   Materials and Methods

RFMirTarget is trained upon biologically validated miRNA-target pairs. This data set is processed by miRanda (see Section 2.3) in order to identify interacting sites between miRNAs and their respective targets and prepare the data set for feature extraction. The alignments provided by miRanda are the source for features definition, which in turn are used for training a random forest classifier. Once trained, the model can be applied to the classification of unknown instances of miRNA-targets (Fig. 1(b)). To perform such task, the user must provide two files with miRNAs and candidate reference sequences, and RFMirTarget outputs the prediction confidence and alignment for all predicted miRNA-target pairs. In what follows we explain the methods involved in the training process, summarized in Fig. 1(a).

2.1   Random Forest

Random forest (RF) is a well-known ensemble approach for classification tasks [17], which has its basis on the combination of tree-structured classifiers with the randomness and robustness provided by bagging and random feature selection. In [18], the application of bagging as a means to enhance the performance of tree-structured classifiers and reduce their bias was proposed by Breiman. The author's approach consisted in training several classifiers with random bootstrap samples from the original data set and afterwards combining their results into a single prediction: for classification tasks, by means of voting; for regression tasks, by averaging all classifiers results.

Furthermore, Breiman has improved his previous model by aggregating random feature selection to the training process [17]. His proposal consisted in selecting from a random subset of features the one with the smallest impurity to split at each node when growing a tree. Tests run by Breiman have revealed that RF classifier always outperforms the bagging approach [18]. However, the benefits of RFs go beyond its good performance. The mechanism applied for growing trees allows an unbiased estimation of both the generalization error and the most important variables for classification during the growth process, using for such analysis the data left out of the bootstrap sample used as training set, named out-of-bag (OOB) data. Additionally, as RFs are tree-structure classifiers, they inherit the interpretability associated to this type of model [9]. In the present work, the RF model was implemented with the `randomForest` R package [19].

2.2   Data Set

The RF model was trained with experimentally verified examples of human miRNA-target collected by Bandyopadhyay and Mitra in [20] and used in the training process of MultiMiTar [16], a SVM-based miRNA-target prediction system. The data set is composed of 289 biologically validated positive examples and 289 systematically identified tissue-specific negative examples. As the basic mechanism of RF renders the definition of training and testing data sets unnecessary, both classifier model and error estimative are drawn from the same data set. During the training process about 2/3 of the original data is sampled for growing the tree, while the remainder (OOB data) is used to test the generated model and estimate the generalization error. We refer the reader to [20] for more details about the data.

2.3   Data Preparation

The data set of positive and negative examples of miRNA-target pairs used for training MultiMiTar does not comprises information about the actual site of alignment between miRNAs and their targets. Due to the existence of multiple sites, which is related to some extent to the short length of miRNAs sequences, the extraction of this information by techniques such as BLAST could result in an extremely large data set, with many biologically unlikely miRNA:mRNA

```
                        20                            1
hsa-let-7a:  3' ttGATATGTTGGATGATGGAGt 5'
                |  |  |||  |||||||||||
HMGA2:  5' atCAAAACACACTACTACCTCt 3'
```

Fig. 2. Example of miRNA-target alignment predicted by miRanda. The miRanda software outputs all possible alignments and base pairings of a miRNA-target pair that scored above a given threshold. The highlighted nucleotides refer to the seed region. In addition, the figure illustrates the nucleotides numbering (1–20) for position-based features extraction.

pairs. Therefore, to reduce the dimension of our problem and prepare our data set for feature extraction, we resort to miRanda software [21] to obtain the miRNA-target binding sites from the same examples used for MultiMiTar training.

The miRanda software runs a score-based algorithm to analyze the complementarity of nucleotides (A:U or G:C) between aligned sequences. The scoring matrix allows the occurrence of the non-canonical base-pairing G=U wobble, which is important for the accurate detection of RNA:RNA duplexes, and is based on the following parameters: +5 for G≡C, +5 for A=U, +2 for G=U and -3 for all other nucleotides pairing [21]. Additionally, there is a scaling factor for giving a higher weight to nucleotides within the first eleven positions.

Besides the scoring matrix, four empirical rules are applied for the identification of the miRNA binding sites, counting from the first position of the 5' end of the miRNA: i) no mismatches at positions 2 to 4; ii) fewer than five mismatches between positions 3-12; iii) at least one mismatch between positions 9 and L-5 (where L is length of the complete alignment); and iv) fewer than two mismatches in the last five positions of the alignment [21]. An example of the alignment output provided by miRanda is depicted in Fig. 2. After running miRanda algorithm on the data set described at Section 2.2, we obtained 1074 positive and 407 negative miRNA-target pairs, which consists of the training instances used for building the RF model with the **randomForest** R package [19]. At this point we emphasize that although the number of training instances we use is greater than the value reported in [16], they derive from the original data set used for training MultiMitar. The difference in the data set dimension is due to data processing by miRanda, which is part of our strategy to fulfill a lack of information on the actual binding site between miRNA-target pairs.

2.4   Features

The negative and positive examples predicted by miRanda algorithm consist of the alignment between both sequences, as well as properties such as the score and alignment length, based on which the classifier features are extracted. The RF features are divided into four categories: structural features, thermodynamics features, alignment features and position-based features. The first two categories are widely used for training classifier systems, while the later was introduced in the study held by [22] and the alignment features are proposed in the current study. In what follows we explain each of the defined categories.

Fig. 3. RFMirTarget combines 34 features divided into four categories: alignment, thermodynamics, position-based and structural features. From these, six refer to the seed region: MFE value and five structural features.

- **Structural features.** Quantify the number of matches (G:C and A:U pairing) and mismatches (G:U wobble pair, gap and other mismatches) in the alignment.
- **Thermodynamics feature.** Minimum free energy (MFE) of the miRNA-target alignment computed by RNAfold [23].
- **Alignment features.** Properties of the miRNA-target alignment computed by the miRanda algorithm: alignment score and alignment length.
- **Position-based features.** Evaluation of each basepair from the 5'-most position of the miRNA up to the 20th position of the alignment, assigning nominal values to designate the kind of pairing in each position: a G:C match, an A:U match, a G:U wobble pair, a gap and a mismatch. In the previous study [22], the gap feature was not included in the set of position-based features.

The thermodynamics and structural features were extracted in twofold manner: for the complete alignment and for the seed region, which is composed by the nucleotides in positions 2-8, to count from the 5'-most position of the miRNA (Fig. 2). Following the approach in [22], the MFE for both seed and complete alignment was computed by using a linker sequence to connect the miRNA and target sequences into a single linear sequence and make possible the use of the RNAfold program (which requires a single linear RNA sequences as input). In the present work we used the same linker sequence applied in [22], "AAAGGGLL-LLLCCCUUU", which according to authors ensures that each part of the subsequence extracted from the alignment will be paired and does not change the thermodynamics qualitatively. In total, 34 features were drawn from the miRanda output, six of which referring to thermodynamics and structural features of the seed region and the remainder concerning properties of the complete alignment, as shown in Fig. 3.

2.5   Performance Assessment

The tools's performance was assessed by computing the total prediction accuracy (ACC), specificity (SPE), sensitivity (SEN) and Matthew's correlation coefficient (MCC) based on the confusion matrix. This matrix is provided by the training process and quantifies the number of instances from the OOB data classified as false positive (FP), true positive (TP), false negative (FN) and true negative (TN).

Table 1. Confusion matrix for RFMirTarget trained upon 28 features

		Prediction		
		Non-Target	Target	Error
Real	Non-Target	346	61	0.1498
	Target	25	1049	0.0232

$$ACC = \frac{TP + TN}{TP + TN + FP + FN} \tag{1}$$

$$SPE = \frac{TN}{TN + FP} \times 100\% \quad (2) \qquad\qquad SEN = \frac{TP}{TP + FN} \times 100\% \quad (3)$$

$$MCC = \frac{TP \times TN - FP \times FN}{\sqrt{(TP + FP) \times (TN + FN) \times (TP + FN) \times (TN + FP)}} \tag{4}$$

3   Results

3.1   RFMirTarget Prediction Performance Based on 28 Features

At first, the RF classifier was trained with a set of 28 features, which comprises all the features summarized in Fig. 3 except for those concerning the seed region, i.e., the MFE and five structure-based features that quantify the number of matches and mismatches between nucleotides 2-8 (see Fig. 2). We adopted the standard values suggested by the `randomForest` R package and trained the model with 500 trees and 5 predictors each. The number of predictors is computed as the square root of the number of variables and this formula is known to lead to results near optimal values [19].

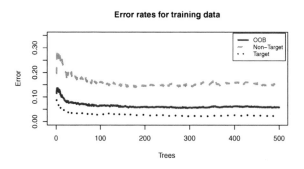

Fig. 4. Error rates for RFMirTarget trained with 28 features. The generalization error decreases as the number of trees in the ensemble prediction increases.

Classification results in terms of the confusion matrix are reported in Table 1. RFMirTarget is very accurate, specially in what concerns the positive examples: the classification error for the target class was minimal, close to 0.02. From the

1074 positive examples, only 25 were misclassified, indicating an outstanding performance to detect true positive targets. The classification error for the non-target class was higher, but still satisfactory: 346 out of the 407 negative examples were correctly classified. The performance metrics for the 28-features model are ACC: 94.19, SEN: 97.67, SPE: 85.01 and MCC: 0.85.

Fig. 4 depicts the evolution of the error rate for the target class, non-target class and OOB data according to the number of trees used in the prediction. Error values start to stabilize from 200 trees. Yet, experiments have shown that there is still a performance gain when adopting 500 trees. In addition, the advantage of using ensemble predictions is clear: the error for the predictions based on 500 trees is much lower than the predictions of a single tree for all three cases.

As previously mentioned in Section 2.1, one important utility of RF classifiers is that they naturally provide an estimative of features importance computed as the forest building progresses. We analyzed this information and found that seven out of the ten most important features according to the average decrease in accuracy (Fig. 5) refer to the seed region, more specifically to the position-based features regarding the seed location (Pos_2 to Pos_8). Hence, we decided to extend our model and include the seed thermodynamics and structural features, training a new RF classifier with the complete set of 34 features summarized in Fig. 3.

3.2   RFMirTarget Prediction Performance Based on 34 Features

The RF model trained with the set of 34 features comprising the complete alignment and seed properties have presented moderated improvement. Results are summarized in Table 2. Compared to the case with 28 features, despite the slight increase in the error rate for the target class, a significant decrease of 31% was observed in the misclassified negative examples: the error rate reduced to 0.10. An analysis of Table 2 in contrast to Table 1 shows that the overall error rate of RFMirTarget suffered a slight reduction, from 0.058 to 0.051. Fig. 6 illustrates the variation in the error rates for the target class, non-target class and OOB

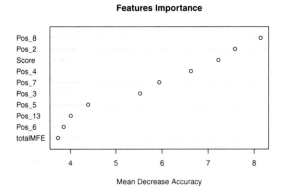

Fig. 5. Features importance for the 28-features RFMirTarget model. Features related to seed region play a crucial role in prediction accuracy.

Table 2. Confusion matrix for the 34-features RFMirTarget model.

		Prediction		
		Non-Target	Target	Error
Real	Non-Target	365	42	0.1031
	Target	35	1039	0.0325

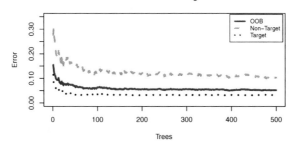

Error rates for training data

Fig. 6. Error rates for RFMirTarget trained upon 34 features. Again, the ensemble prediction provides a more accurate result than single classifier prediction.

data according to the number of trees used in the training process: the error rates decrease as the number of trees increase.

A comparison between the 28-features and 34-features models in terms of the performance metrics discussed in Section 2.5 is given in Table 3. We include also the performance values reported for MultiMiTar [16]. A slight performance gain is observed for the 34-features model in relation to the 28-features model regarding all metrics except sensitivity. The greatest increase is related to the specificity, which suggests that information about the seed region is important to the correct identification of true negative instances, and thus to decrease the rate of false positives. Moreover, the analysis of the 34-features model results in contrast to MultiMiTar indicates that RFMirTarget has a balance between specificity and sensitivity as good as the latter. In addition, RFMirTarget shows an increase of 48% and 18.5% in MCC and accuracy, respectively, when compared to MultiMitar. Despite the fact that MultiMiTar performance metrics are computed based on an independent testing data set, the confusion matrix generated by RFMirTarget also provides an unbiased measurement of the performance [17] due to the use of bootstrap samples when training the RF model. Hence, this process generates a reliable method assessment.

The analysis of the ten most important features revealed that the features related to the seed region have the greatest impact on method's performance, as expected (Fig. 7). According to this analysis, the most relevant features are the MFE value and the number of GC base pairs of the seed region, which is biologically plausible since GC pairings are more stable because they involve three hydrogen bonds.

Table 3. Comparison between RFMirTarget models trained with 28 features and 34 features, in contrast to the competing tool MultiMiTar. All models are trained upon the same data set. *Values reported in the tools' original paper [16].

	ACC (%)	SEN (%)	SPE (%)	MCC
RFMirTarget, 28-features model	94.19	97.67	85.01	0.85
RFMirTarget, 34-features model	94.80	96.74	86.68	0.86
MultiMiTar*	80.00	89.83	70.21	0.58

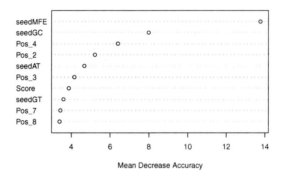

Fig. 7. Features importance for the 34-features RFMirTarget model. Besides the score, which is associated to the complete alignment, all relevant features concern the seed region. The MFE value of the seed region, in particular, causes a high impact in the prediction accuracy.

4   Conclusion and Future Work

In the present paper we introduced RFMirTarget, a classifying system for miRNA targets prediction based on the RF algorithm. In [9], RFs were introduced as a tool for predicting miRNA precursors and identifying novel miRNAs, performing better than the well-known SVM classifier. Moreover, RF algorithms are of easy implementation, require less computational resources and time, and they are of more easy understanding for the final user when compared to other classification methods such as SVM. As RFs are ensembles of classification trees, they inherit the interpretability property of the latter and can be easily translated into rules, hence representing an interesting tool for exploring data [24]. Nonetheless, none of the tools for miRNA-target prediction proposed so far were built upon RFs.

The first RF model presented in the current work was trained with 28 structural, thermodynamics, alignment and position-based features extracted from the complete alignment between miRNA and candidate target. The results were encouraging: 94% of the examples were correctly classified. Seven out of the ten most important features are related to seed properties. Aggregating six thermodynamics and structural features of the seed region in the training process of the RF model, the classification performance improved in almost all senses. The list of the ten most important features for this model includes only seed features, except for the alignment score provided by the miRanda. Thus, our tool reinforces the importance of the seed region for target prediction accuracy, corroborating previous studies in the area [13,14].

Comparing RFMirTarget with the competing tool MultiMiTar [16], which has presented the best predictions results for miRNAs targets so far, RFMirTarget has superior performance in all aspects. The MCC was significantly higher for the RFMirTarget: 0.86 against 0.58, which represents an improvement of 48%. The accuracy was also enhanced nearly to 18.5%: RFMirTarget has correctly predicted 94.80% of the examples, in contrast to 80.00% of MultiMiTar. MultiMiTar is a SVM classifier trained upon the same data set used in this work and its performance on an independent testing data set is superior than former methods [16]. Therefore, one can conclude that RFMirTarget is a reliable and robust strategy for miRNA target prediction when compared to other existing popular methods.

An interesting direction for future work is a deeper investigation of the impact of seed region for miRNAs targets prediction. Features related to the seed region are predominant in the analysis of the most impacting features regarding model's accuracy, such that we find worth investigating the predictive power of a classifying model trained based on this set of features, and perform comparisons with other available tools using an independent testing data set. RFMirTarget should be soon made freely available for download under a GPL license.

References

1. Bartel, D.P.: MicroRNAs: Genomics, review biogenesis, mechanism, and function. Cell 116, 281–297 (2004)
2. Betancur, J.G., Tomari, Y.: Dicer is dispensable for asymmetric RISC loading in mammals. RNA 18(1), 1–7 (2011)
3. Lee, R.C., Feinbaum, R.L., Ambrost, V.: The C. elegans Heterochronic Gene lin-4 Encodes Small RNAs with Antisense Complementarity to lin-14. Cell 75, 843–854 (1993)
4. Lu, M., Zhang, Q., Deng, M., Miao, J., Guo, Y., Gao, W., Cui, Q.: An analysis of Human microRNA and disease associations. PLoS ONE 3(10), e3420 (2008)
5. Chen, X.: microRNA biogenesis and function in plants. FEBS Letters 579, 5923–5931 (2005)
6. Liu, J., Zheng, M., Ling Tang, Y., Hua Liang, X., Yang, Q.: microRNAs, an active and versatile group in cancers. Int. J. Oral. Sci. 3, 165–175 (2011)
7. Xue, C., Li, F., He, T., Liu, G.P., Li, Y., Zhang, X.: Classification of real and pseudo microRNA precursors using local structure-sequence features and support vector machine. BMC Bioinformatics 6(1), 310 (2005)
8. Batuwita, R., Palade, V.: microPred: effective classification of pre-miRNAs for human miRNA gene prediction. Bioinformatics 25(8), 989–995 (2009)
9. Jiang, P., Wu, H., Wang, W., Ma, W., Sun, X., Lu, Z.: MiPred: classification of real and pseudo microRNA precursors using random forest prediction model with combined features. Nucleic Acids Research 35, 339–344 (2007)

10. Yousef, M., Nebozhyn, M., Shatkay, H., Kanterakis, S., Showe, L.C., Showe, M.K.: Combining multi-species genomic data for microRNA identification using a Naïve Bayes classifier. Bioinformatics 22(11), 1325–1334 (2006)
11. Zhang, Y.: miRU: an automated plant miRNA target prediction server. Nucleic Acids Research 33, W701–W704 (2007)
12. Lytle, J.R., Yario, T.A., Steitz, J.A.: Target mRNAs are repressed as efficiently by microRNA-binding sites in the 5' UTR as in the 3' UTR. PNAS 104(23), 9667–9672 (2007)
13. Maziére, P., Enright, A.J.: Prediction of microRNA targets. Drug Discovery Today 12(11/12), 452–458 (2007)
14. Lhakhang, T.W., Chaudhry, M.A.: Current approaches to microRNA analysis and target gene prediction. Journal of Applied Genetics, 1–10 (2011)
15. Witkos, T.M., Koscianska, E., Krzyzosiak, W.J.: Practical aspects of microRNA target prediction. Current Molecular Medicine 11, 93–109 (2011)
16. Mitra, R., Bandyopadhyay, S.: MultiMiTar: A novel multi objective optimization based miRNA-target prediction method. PLoS ONE 6(9), e24583 (2011)
17. Breiman, L.: Random forests. Machine Learning 45(1), 5–32 (2001)
18. Breiman, L.: Bagging predictors. Machine Learning 24(2), 123–140 (1996)
19. Liaw, A., Wiener, M.: Classification and Regression by randomForest. R News 2(3), 18–22 (2002)
20. Bandyopadhyay, S., Mitra, R.: TargetMiner: microRNA target prediction with systematic identification of tissue-specific negative examples. Bioinformatics 25(20), 2625–2631 (2009)
21. Enright, A., John, B., Gaul, U., Tuschl, T., Sander, C., Marks, D.: MicroRNA targets in drosophila. Genome Biology 5(1), R1 (2003)
22. Kim, S.K., Nam, J.W., Rhee, J.K., Lee, W.J., Zhang, B.T.: Mitarget: microRNA target gene prediction using a support vector machine. BMC Bioinformatics 7(1), 411 (2006)
23. Hofacker, I.L.: Vienna RNA secondary structure server. Nucleic Acids Research 31(1), 3429–3431 (2003)
24. Tkacz, A., Rychlewski, L., Uva, P., Plewczynski, D.: Supervised classification of genes and biological samples. In: de Rinaldis, E., Lahm, A. (eds.) DNA Microarrays: Current Applications, 1st edn., pp. 101–120. Taylor & Francis (2007)

Prediction of Transcription Factor Binding Sites by Integrating DNase Digestion and Histone Modification

Eduardo G. Gusmão[1], Christoph Dieterich[2], and Ivan G. Costa[1]

[1] Center of Informatics, Federal University of Pernambuco, Recife, Brazil
{egg,igcf}@cin.ufpe.br
[2] Berlin Institute for Medical Systems Biology, Berlin, Germany
christoph.dieterich@mdc-berlin.de

Abstract. The identification of cis-acting elements on DNA is crucial for the understanding of the complex regulatory networks that govern many cell mechanisms. However, this task is very complex since it is estimated that there are 1500 different transcription factors (TFs) in the human genome, each of which can bind to multiple loci directly or indirectly. The standard computational approach is the use of a position weight matrix (PWM) to represent the binding preference of a transcription factor and the use of statistical procedures to detect genomic regions with high binding scores. Given the small and degenerate signals of most PWMs, such approach suffers from a very high number of false positive hits. Current research has proven that genome wide assays reflecting open chromatin, such as DNase digestion or histone modifications, can improve sequence based detection of the binding location of transcription factors that are active in a particular cell type. We propose here a Multivariate Hidden Markov Model that is able to improve the prediction of transcription factor binding locations by integrating DNase digestion and histone modification data. Our methodology improves sensitivity, in comparison to existing methods, with little or no effect at specificity rates. This study shows that it is possible to improve predictability power of cis-acting elements by correctly integrating DNase and histone modification data, allowing for more sophisticated studies using a larger set of epigenetic signals.

Keywords: cis-regulatory elements, DNase I-hypersensitive sites, histone modifications, hidden markov models.

1 Introduction

Complex regulatory networks govern many critical cell mechanisms such as proliferation, development, differentiation, aging and apoptosis. The regulatory mechanism consists of a large number of different components that may play a role in numerous regulatory pathways [1]. Examples of these components are trans-acting elements (or transcription factors), cis-acting elements (such as enhancers, silencers and insulators) and epigenetic factors (such as histone

M.C.P. de Souto and M.G. Kann (Eds.): BSB 2012, LNBI 7409, pp. 109–119, 2012.

modifications, chromatin remodeling complexes and DNA methylation), each collaborating in the orchestration of proper temporal and spatial expression required by ubiquitous, common or cell-specific processes [2]. Consequently, it is crucial to identify these regulatory elements in order to understand their role in each cell's regulatory network and comprehend diseases caused by deregulation.

The identification of cis-acting elements, which transcription factors bind to, can be a particularly daunting task since it is estimated that there are 1500 factors in the human genome [3]. Furthermore, transcription factor binding sites (TFBSs) are generally small, in the range of 6–12 bp with binding specificity dictated by no more than 4–6 positions within this site [2], and only a subset of them are active during a current cell state [4]. In addition, many transcription factors may have multiple binding sites with different motifs [2]; and some of these elements may bind to the DNA by indirectly binding with another factor or protein complex. The standard computational approach — motif matching — uses position weight matrix (PWM) representation of transcription factor binding preference followed by a statistical procedure to detect genomic regions with a high binding score for a particular TF [5]. Nevertheless, motif matching is highly dependent on the strength of the matching algorithm, it can not distinguish between active and non-active binding sites and presents a high number of false positives as motifs are mostly small and degenerate [2]. An alternative are Chromatin immunoprecipitation assays, which can be applied to finding binding sites on a genome-wide fashion (ChIP-Seq). However such experiments are condition-specific, fails for some particular TFs and are experimentally and financially demanding [6].

Current research has proven that genome wide assays reflecting open chromatin states, such as DNase digestion data (DNase-Seq) [3,4] or histone modifications (ChIP-Seq) [7], can improve sequence based detection of the binding location of transcription factors that are active in a particular cell type. The rational of such approaches is to restrict the sequence based search of binding sites to genomic regions where either DNase digestion or histone marks indicates the chromatin is open and accessible for TF binding. Traditional open chromatin assays with DNase I as cleavage agent have long been used to characterize hypersensitive sites, i.e. genomic regions with many sites digested by DNase, and it is considered a high-accuracy and high-resolution technique [8]. This technique can be combined with high-throughput sequencing to provide genome-wide maps of open chromatin of a particular cell type [9]. Similarly, ChIP-Seq assays can be used to measure patterns of histone variants and their post-translational modifications (such as acetylation and methylation) in different cell lines. Many studies have shown clear chromatin signatures for particular regulatory regions, such as active promoters and enhancers, and have suggested the application of their findings to predict elements of the regulatory mechanism [4,7,10].

We propose here a multivariate Hidden Markov Model (HMM) that is able to identify transcription factor binding sites (TFBSs) by integrating DNase digestion data and histone modification data of a particular cell type. The method outperforms previous models based on HMMs, which made predictions either with histone modification [7] or DNase digestion [3] alone. Moreover, we also propose here a procedure to estimate the HMM parameters without necessity of traditional DNase digestion data as in [3]. We evaluate the method by using DNase digestion data and histone modifications from the leukemia cell line K562 to predict TFBSs of four TFs: ATF3, CTCF, GABP and REST.

1.1 Related Work

Boyle et al [3] proposed an HMM model, which used DNase-seq data to predict TFBSs that are active in a particular cell type. The HMM parameters were obtained with a manual annotation of DNase hypersensitivity sites. Won et al [11] used a multivariate HMM model with scores produced by motif matching with PWMs and ChIP-seq signals of histone modifications to predict binding sites of individual transcription factors. Cuellar-Partida et al [4] described a method for combining epigenetic data with standard DNA sequence on a Bayesian approach. Pique-regi et al [12] created an algorithm that combines genomic sequence information (such as conservation) with cell-specific experimental data (such as DNase and histone modifications) using a Bayesian approach.

2 Materials and Methods

The proposed methodology works as follows. As a first step, we perform a simple motif matching experiment to detect sequence based predictions of TFBSs (Section 2.1). From this point, we will use the nomenclature Motif Predicted Binding Sites (MPBSs) to define these sequence based TFBSs candidates. Next, we detect DNase hypersensitive regions of the target cell type (Section 2.2). Our search for TFBSs will be restricted to this region only. Then, high-resolution signal of the DNase digestion and histone modification data from the target cell type are used as input for the HMM (Section 2.3), which was previously trained using manually annotated footprint regions or regions around sequence based TFBS predictions (Section 2.4). The HMM uses DNase digestion and histone modification signals to detect footprint regions, i.e. small regions within DNase hypersensitive sites where TFs are likely to bind. We regard as 'predictions', MPBSs that matche footprint regions. Finally, we use ChIP-seq data for specific TFBSs to validate the accuracy of the HMMs (Section 2.5).

2.1 Motif Matching

All the PWMs were obtained from one of these repositories: Jaspar [13], Transfac [14] and Uniprobe [15] , and a high-quality CTCF motif was obtained from Renlab [16]. A PWM is a representation for sites of preferred binding by specific transcription factors that consists of evaluated levels of affinity for each position of the motif and for each nucleotide. There are some redundancies among these databases, i.e. there are many motifs for the same factor, but as each one has its own quality that depends on the particular study that generated it, we analyzed each one separately. All the motifs were matched against the complete genome using the motif matching tool available in Biopython [17] to produce a bit score. Next, matches (MPBSs) were discarded if they had scores lower than the maximum between 70% of the highest possible bit score and 90% of the difference between the highest possible or the lowest possible bit score [3]. We would like to state that we are aware that more robust motif matching algorithms exist and our intention on using these criteria is to replicate [3] providing a reliable comparative scenario.

2.2   Detection of DNase Hypersensitive Regions

DNase Hypersensitive Regions were detected as described in the ENCODE project [1]. Briefly, the regions of significant enrichment of DNase digestion tags (HS regions) were identified using the F-seq method [18]. This computational tool identifies regions of high density of sequence reads by applying a Parzen window methodology and considering a random background distribution of reads. Then, the resulting distribution of F-seq scores (at bp resolution) was fitted to a gamma distribution and a P-value of 0.05 was used to discretely define the regions more likely to be in open chromatin state. For the cell line K562, there are 133.372 DNase hypersensitivity regions.

2.3   DNase and Histone Modification High-Resolution Signal Generation

We use high-resolution signals of DNase digestion and histone modification as input for our HMMs. For DNase data, we apply the same procedure as proposed in [3]. First, we counted all the 5' bp (corresponding to the exact position at which DNase I enzyme has nicked the DNA) generating a digestion signal at the bp resolution. Then, this signal was normalized by dividing each bp count by the mean of all non-zero entries in a 1kb window surrounding that bp. Finally, the slope of this normalized signal was obtained by the Savitzky-Golay method. In this method, the data are fitted to a second order polynomial and a convolution method is applied to access the first derivative based on a window size of 8bp.

This signal corresponding to the slope of the normalized curve assumes positive values when there is an increase and negative values when there is a decrease. The higher the slope signals are, the steeper the normalized increase is, and the lower the slope signals are, the steeper the decrease is. This strategy is performed to access the DNase footprint patterns previously described, where the transcription factor binding sites are clearly characterized as a depletion in the DNase I digestion between two peaks [3]. Consequently, the slope signal of the read counts permits the training of an HMM that has the necessary states to capture this peak-dip-peak pattern. See Figure 1 rigth for an example of DNase signals.

The histone modification signal was based at the read-counts level. We extended each histone fragment mapped to the genome, which originally contained 36bp, to exactly 200bp. The resulting signal consisted of the natural logarithm of the counted reads. The log step is done to soften the parameter values of the

HMM between different states and make its distribution closer to normal. For this experiments, we have used histone modifications with a known role for indicating active regulatory regions: H3k4me2, H3k4me3, H3k9ac and the histone variant H2A.Z.

2.4   Detection of Footprints with HMMs

A five-state bivariate emission HMM was used to integrate the signal of DNase and one histone modification to predict open chromatin (see Figure 1 left for the HMM topology). The first state (labeled BACK) represents the background of the HS site. Typically, this state recognizes regions of low or moderate DNase digestion and low histone signal. The second state (HH) was created to capture the high histone marks that occur at the surroundings of the DNase HS core (region where there are higher concentrations of peaks). The rest of the model followed the methodology described in [3]. The UP state recognizes increasing DNase digestion peaks while DOWN state recognizes decreasing signals. The footprint state (FP) is characterized then after the DOWN state and before another UP state, according to the peak-dip-peak trend. The most noticeable difference between our bivariate model and Boyle et al's univariate model is that, in our model, the HMM enters in the open chromatin identification states (UP, DOWN and FP) only after a significant increase in the histone modification signal. Moreover, we expect small levels of histone marks at the FP state (see average signals of the DNA digestion and of the H2A.Z histone mark for a 1000bp region centered at the 100 top scored MPBSs at Figure 1 right). The posterior probability was used to determine the most probable state.

We performed two different training approaches. The first one, used by Boyle et al [3], consists of training via maximum-likelihood algorithm an HMM model with a manually annotated DNase hypersensitive region corresponding to the promoter region of the fragile X mental retardation 1 (FMR1) gene [19]. This model is then used to annotate the 1000 top scored HS regions of the chromossome 6, with the assumption that these strong sites would consist of ubiquitous regulatory features. A final HMM model is obtained after re-estimation of parameters with these annotated sites. We will refer to this aproach as FMR1 in the subsequent text. We also developed a novel approach that requires no high resolution initial footprint region. We matched all the motifs in jaspar, transfac and uniprobe repositories against the 10 top scored HS regions using the STAMP algorithm [20]. These sites were then annotated, used to train the HMM model and excluded from further analysis. The General Hidden Markov Model (GHMM) python binding package was used to implement the HMMs [21].

2.5   TFBS Gold Standard

The gold standards were defined as in previous studies [3,4]. We use the TFBS predictions from genome-wide ChIP-Seq of the leukemia cell line K562 provided by ENCODE (data tracks HAIB, SYDH and UTA) for the factors ATF3, CTCF, GABP and REST (see ENCODE project for exact details on TFBS prediction).

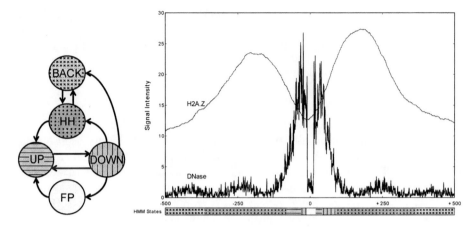

Fig. 1. HMM topology (left) and the mean DNase digestion and H2A.Z histone signal for a 1000bp region centered at the 100 top scored MPBSs within a ChIP-seq enriched region for CTCF (right). We display below an example plot of the most probable HMM states for the mean signals.

After performing the motif matching, the MPBSs that contained ChIP-seq evidence were considered positives and those which did not contain such evidence were considered negatives. Therefore, we can state the following: MPBSs that contained ChIP-seq evidence and were predicted by the model are considered true positives, MPBSs that contained ChIP-seq evidence but were not predicted by the model are false negatives, MPBSs that did not contain ChIP-seq evidence but were identified by the model are false positives and MPBSs that did not contain either ChIP-seq evidence or footprint evidence are considered true negatives. This criterion allows a quantitative measurement of the performance of footprinting algorithms. We calculated the sensitivity (Sn) as $TP/(TP + FN)$, specificity (Sp) as $TN/(TN + FP)$, positive predictive value (Pp) as $TP/(TP + FP)$, negative predictive value (Np) as $TN/(TN + FN)$ and correct rate (Cr) as $(TP + TN)/(TP + FN + TN + FP)$.

3   Results and Discussion

We focused our analysis on four different transcription factors: ATF3 (cyclic AMP-dependent transcription factor), CTCF (CCCTC-binding factor), GABP (GA-binding protein) and REST (or NRSF). We use one ATF3 motif (transfac), two CTCF motifs (jaspar and renlab), one GABP motif (jaspar) and two REST motifs (jaspar and transfac), making a total of 6 motifs. We performed the analyis by combining the DNase digestion with either the histone variant (H2A.Z) or one of the four histone modifications (H3K4me2, H3K4me3 abd H3K9ac). For simplicity, we will refer to both histone variants and histone modifications as histone modifications.

To integrate histone modification into our model, we analyzed their patterns surrounding known TFBSs (Figure 1 (right)). This figure show a clear peak-dip-peak trend for both the H2A.Z and DNase digestion. However, unlike DNase digestion signal, histone modification signal presents a smooth appearance and its dips usually span over longer regions since ChIP-seq methodology provides lower resolution than DNase-seq. Moreover the decrease in histone levels are simultaneous to the increase of the DNase digestion sites. Similar profiles were obtained for other TFs and histone modifications.

After the analysis of the histone signal, eight HMM models were created (with the topology depicted in Fig. 1 left). These models correspond to the FMR1-based training and STAMP-based training of four models, each one composed by DNase-seq data plus one of the histone modifications. As an initial study, we did not create models with dimension larger than 2 as that would increase significantly the chance of overfitting. Also, we wanted to compare the predictive power of each histone modification separately. We then performed the footprinting in the statistically determined HS regions of cell line K562.

The Table 1 exhibits results for the accuracy assessment criteria described in Section 2.5. The statistics are evaluated in order to measure the capability of both DNase only and DNase combined with histone modifications, to predict CTCF motif from jaspar repository.

Table 1. Results on CTCF (jaspar) for all models trained with either FMR1 or STAMP methodologies. For each statistic, the best result is marked in bold.

Model	Train	Sn	Sp	Pp	Np	Cr
DNase only	FMR1	29.45	99.59	**99.87**	11.35	35.28
	STAMP	26.08	**99.86**	99.95	10.91	32.21
H2A.Z	FMR1	50.33	97.93	99.63	15.16	54.29
	STAMP	71.80	94.74	99.34	23.35	73.71
H3K4me2	FMR1	63.71	95.85	99.41	19.32	66.38
	STAMP	74.76	94.74	99.37	25.39	76.42
H3K4me3	FMR1	65.13	96.13	99.46	19.99	67.71
	STAMP	**75.45**	94.33	99.32	**25.83**	**77.02**
H3K9ac	FMR1	60.95	96.68	99.51	18.33	63.92
	STAMP	74.96	94.33	99.32	25.46	76.57

The inclusion of any histone modifications always outperforms the DNase only HMM. The latter, usually achieves high specificity but low sensitivity. The inclusion of histone modifications improves drastically the sensitivity with a small decrease in specificity. Moreover, the STAMP training performed better than FMR1 training when histone signal were included. All models achieved around 75% of correct predictions, increasing about 40% in sensitivity while losing, at most, 5% in specificity.

Table 2. Results on ATF3, CTCF, GABP and REST factors, for DNase only model trained with FMR1 and our model trained with STAMP. For each statistic, the best result(s) is(are) marked in bold.

		DNase only	New Model			
			H2A.Z	H3K4me2	H3K4me3	H3K9ac
ATF3 (Transfac)	Sn	58.75	70.00	76.25	**77.50**	67.50
	Sp	**96.80**	90.32	89.68	89.32	89.88
	Pp	**10.28**	4.31	4.41	4.33	3.99
	Np	99.73	99.79	**99.84**	**99.84**	99.77
	Cr	**96.57**	90.19	89.60	89.25	89.74
CTCF (Ren)	Sn	29.68	69.16	72.51	**72.98**	72.27
	Sp	**98.40**	88.26	87.82	87.45	88.11
	Pp	**98.82**	96.38	96.42	96.34	96.49
	Np	23.64	38.77	41.42	**41.73**	41.29
	Cr	42.13	72.62	75.29	**75.60**	75.14
GABP (Jaspar)	Sn	27.90	46.32	50.87	**53.11**	42.19
	Sp	**99.77**	94.96	94.55	94.37	94.49
	Pp	**91.84**	46.27	46.61	46.91	41.74
	Np	93.66	94.97	95.36	**95.56**	94.58
	Cr	**93.62**	90.80	90.81	90.84	90.01
REST (Jaspar)	Sn	20.49	**55.21**	55.04	55.04	**55.21**
	Sp	**96.67**	95.00	95.00	95.00	95.00
	Pp	99.18	**99.54**	**99.54**	**99.54**	**99.54**
	Np	5.82	**9.73**	9.69	9.69	**9.73**
	Cr	24.17	**57.13**	56.97	56.97	**57.13**
REST (Transfac)	Sn	31.78	**69.44**	68.95	69.19	**69.44**
	Sp	100.0	100.0	100.0	100.0	100.0
	Pp	100.0	100.0	100.0	100.0	100.0
	Np	3.13	**6.72**	6.62	6.67	**6.72**
	Cr	33.25	**70.10**	69.62	69.86	**70.10**

Table 2 allows for a wider range of comparisons. For renlab's CTCF motif and for both REST motifs, the inclusion of histone modification data have shown considerably better results than DNase only.

We must notice that, although our models have increased the sensitivity for ATF3 and GABP, the correct rate was lower than DNase only model. This happened due to the great quantity of negative instances, i.e. MPBSs without ChIP-seq evidence. However, the use of histone modification data presented better results at average, while the accuracy of DNase only model was highly dependent of the factor. Another relevant aspect is the fact that histones modifications have overall close results with H3K4me3, H3k9ac and H2A.Z as best results for distinct factors. These histone modifications, which are well known to be related to active promoter regions, can be alternatively used for the task of predicting active TFBSs.

4   Final Remarks

In this work, we proposed a novel multivariate emission HMM method to predict TFBSs based on DNase I digestion patterns and histone modification data. We analyzed the patterns of histone modifications on various factors and determined the modification types that best suit our interests in this exploratory study. We compared our model with a previously suggested model that uses only DNase digestion as input and demonstrated that, in most cases, ours is as good as or better. This study showed that it is possible to combine different epigenetic signals to improve the precise identification of TFBSs without losing much specificity.

In future studies, the number of cell lines, histone modifications and tested factors will be increased. An initial plan would consist of extending the bivariate model to more than two dimensions. However, this has to be done using more careful strategies since the chance of overfitting is significantly higher. In addition, the model can be created with other types of epigenetic data or even genomic data, that are not cell-type specific (such as sequence conservation or PWM motif matching score). We also plan to analyze the accuracies of our model considering epigenetic relationships between histone modifications and cis-acting regulatory element type (enhancer, silencer, insulator and others). Furthermore, we intent to keep researching training methodologies that do not require laborious manual annotations and are capable of describing general and specific patterns in data. Finally, we plan to investigate more thoroughly which epigenetic signals are better predictors of TFBSs given a particular cell line and factor and draw conclusions about the generalization capability of these epigenetic features.

Acknowledgments. We would like to thank Marcílio C.P. de Souto and Thaís G. do Rêgo for sharing some valuable information and for making helpful suggestions.

Funding. This work has been partially supported by Brazilian research agencies: FACEPE, CNPq and CAPES.

References

1. Rosenbloom, K.R., Dreszer, T.R., Long, J.C., Malladi, V.S., Sloan, C.A., Raney, B.J., Cline, M.S., Karolchik, D., Barber, G.P., Clawson, H., Diekhans, M., Fujita, P.A., Goldman, M., Gravell, R.C., Harte, R.A., Hinrichs, A.S., Kirkup, V.M., Kuhn, R.M., Learned, K., Maddren, M., Meyer, L.R., Pohl, A., Rhead, B., Wong, M.C., Zweig, A.S., Haussler, D., Kent, W.J.: ENCODE Whole-Genome Data in the UCSC Genome Browser: Update 2012. Nucleic Acids Res. 40(Database issue), D912–D917 (2012)

2. Maston, G.A., Evans, S.K., Green, M.R.: Transcriptional Regulatory Elements in the Human Genome. Annu. Rev. Genomics Hum. Genet. 7, 29–59 (2006)
3. Boyle, A.P., Song, L., Lee, B.K., London, D., Keefe, D., Birney, E., Iyer, V.R., Crawford, G.E., Furey, T.S.: High-Resolution Genome-Wide In Vivo Footprinting of Diverse Transcription Factors in Human Cells. Genome Res. Biol. 21(3), 456–464 (2011)
4. Cuellar-Partida, G., Buske, F.A., McLeay, R.C., Whitington, T., Noble, W.S., Bailey, T.L.: Epigenetic Priors for Identifying Active Transcription Factor Binding Sites. Bioinformatics 28(1), 56–62 (2012)
5. Stormo, G.D.: DNA Binding Sites: Representation and Discovery. Bioinformatics 16(1), 16–23 (2000)
6. Park, P.J.: ChIP-seq: Advantages and Challenges of a Maturing Technology. Nature Reviews Genetics 10(10), 669–680 (2009)
7. Hon, G., Wang, W., Ren, B.: Discovery and Annotation of Functional Chromatin Signatures in the Human Genome. PLoS Computational Biology 5(11), e1000566 (2009)
8. Gross, D.S., Garrard, W.T.: Nuclease Hypersensitive Sites in Chromatin. Ann. Rev. Biochem. 57, 159–197 (1988)
9. Crawford, G.E., Holt, I.E., Mullikin, J.C., Tai, D., National Institutes of Health Intramural Sequencing Center, Green, E.D., Wolfsberg, T.G., Collins, F.S.: Identifying Gene Regulatory Elements by Genome-Wide Recovery of DNase Hypersensitive Sites. PNAS 101(4), 992–997 (2004)
10. Barski, A., Cuddapah, S., Cui, K., Roh, T., Schones, D.E., Wang, Z., Wei, G., Chepelev, I., Zhao, K.: High-Resolution Profiling of Histone Methylations in the Human Genome. Cell 129(4), 823–837 (2007)
11. Won, K., Ren, B., Wang, W.: Genome-Wide Prediction of Transcription Factor Binding Sites Using an Integrated Model. Genome Biology 11(1), R7 (2010)
12. Pique-Regi, R., Degner, J.F., Pai, A.A., Gaffney, D.J., Gilad, Y., Pritchard, J.K.: Accurate Inference of Transcription Factor Binding from DNA Sequence and Chromatin Accessibility Data. Genome Res. 21(3), 447–455 (2011)
13. Byrne, J.C., Valen, E., Tang, M.E., Marstrand, T., Winther, O., Piedade, I., Krogh, A., Lenhard, B., Sandelin, A.: JASPAR, the Open Access Database of Transcription Factor-Binding Profiles: New Content and Tools in the 2008 Update. Nucleic Acids Research 36(Database issue), D102–D106 (2008)
14. Matys, V., Kel-Margoulis, O.V., Fricke, E., Liebich, I., Land, S., Barre-Dirrie, A., Reuter, I., Chekmenev, D., Krull, M., Hornischer, K., Voss, N., Stegmaier, P., Lewicki-Potapov, B., Saxel, H., Kel, A.E., Wingender, E.: TRANSFAC and its Module TRANSCompel: Transcriptional Gene Regulation in Eukaryotes. Nucleic Acids Research 34(Database issue), D108–D110 (2006)
15. Newburger, D.E., Bulyk, M.L.: UniPROBE: An Online Database of Protein Binding Microarray Data on Protein? DNA Interactions. Nucleic Acids Research 37(Database issue), D77–D82 (2009)
16. Kim, T.H., Abdullaev, Z.K., Smith, A.D., Ching, K.A., Loukinov, D.I., Green, R.D., Zhang, M.Q., Lobanenkov, V.V., Ren, B.: Analysis of the Vertebrate Insulator Protein CTCF-Binding Sites in the Human Genome. Cell 128(6), 1231–1245 (2007)
17. Cock, P.J.A., Antao, T., Chang, J.T., Chapman, B.A., Cox, C.J., Dalke, A., Friedberg, I., Hamelryck, T., Kauff, F., Wilczynski, B., de Hoon, M.J.L.: Biopython: Freely Available Python Tools for Computational Molecular Biology and Bioinformatics. Bioinformatics 25(11), 1422–1423 (2009)

18. Boyle, A.P., Guinney, J., Crawford, G.E., Furey, T.S.: F-Seq: A Feature Density Estimator for High-Throughput Sequence Tags. Bioinformatics 24(21), 2537–2538 (2008)
19. Drouin, R., Angers, M., Dallaire, N., Rose, T.M., Khandjian, E.W., Rousseau, F.: Structural and Functional Characterization of the Human FMR1 Promoter Reveals Similarities with the hnRNP-A2 Promoter Region. Human Molecular Genetics 6(12), 2051–2060 (1997)
20. Mahony, S., Benos, P.V.: STAMP: A Web Tool for Exploring DNA-Binding Motif Similarities. Nucleic Acids Research 35(Web Server issue), W253–W258 (2007)
21. The General Hidden Markov Model Library (GHMM), `http://ghmm.org/`
22. Boyle, A.P., Davis, S., Shulha, H.P., Meltzer, P., Margulies, E.H., Weng, Z., Furey, T.S., Crawford, G.E.: High-Resolution Mapping and Characterization of Open Chromatin across the Genome. Cell 132(2), 311–322 (2008)
23. Crawford, G.E., Davis, S., Scacheri, P.C., Renaud, G., Halawi, M.J., Erdos, M.R., Green, R., Meltzer, P.S., Wolfsberg, T.G., Collins, F.S.: DNase-chip: A High Resolution Method to Identify DNase I Hypersensitive Sites Using Tiled Microarrays. Nature Methods 3(7), 503–509 (2006)
24. Crawford, G.E., Holt, I.E., Whittle, J., Webb, B.D., Tai, D., Davis, S., Margulies, E.H., Chen, Y., Bernat, J.A., Ginsburg, D., Zhou, D., Luo, S., Vasicek, T.J., Daly, M.J., Wolfsberg, T.G., Collins, F.S.: Genome-Wide Mapping of DNase Hypersensitive Sites Using Massively Parallel Signature Sequencing (MPSS). Genome Res. 16(1), 123–131 (2006)
25. Zhang, Y., Liu, T., Meyer, C.A., Eeckhoute, J., Johnson, D.S., Bernstein, B.E., Nusbaum, C., Myers, R.M., Brown, M., Li, W., Liu, X.S.: Model-Based Analysis of ChIP-seq (MACS). Genome Biology 9(9), R137.1–R137.9 (2008)

Evaluating Correlation Coefficients for Clustering Gene Expression Profiles of Cancer

Pablo A. Jaskowiak[1], Ricardo J.G.B. Campello[1], and Ivan G. Costa[2]

[1] Department of Computer Sciences
Institute of Mathematics and Computer Sciences
University of São Paulo - São Carlos, Brazil
{pablo,campello}@icmc.usp.br
[2] Center of Informatics
Federal University of Pernambuco - Recife, Brazil
igcf@cin.ufpe.br

Abstract. Cluster analysis is usually the first step adopted to unveil information from gene expression data. One of its common applications is the clustering of cancer samples, associated with the detection of previously unknown cancer subtypes. Although guidelines have been established concerning the choice of appropriate clustering algorithms, little attention has been given to the subject of proximity measures. Whereas the Pearson correlation coefficient appears as the *de facto* proximity measure in this scenario, no comprehensive study analyzing other correlation coefficients as alternatives to it has been conducted. Considering such facts, we evaluated five correlation coefficients (along with Euclidean distance) regarding the clustering of cancer samples. Our evaluation was conducted on 35 publicly available datasets covering both (i) intrinsic separation ability and (ii) clustering predictive ability of the correlation coefficients. Our results support that correlation coefficients rarely considered in the gene expression literature may provide competitive results to more generally employed ones. Finally, we show that a recently introduced measure arises as a promising alternative to the commonly employed Pearson, providing competitive and even superior results to it.

Keywords: correlation, proximity measure, clustering, gene expression.

1 Introduction

Microarray technology enables expression level measurement for thousands of genes in a parallel fashion, helping researchers to gather knowledge and insight about diverse biological phenomena. In order to unveil information contained in gene expression data, one of the first steps usually adopted is cluster analysis, which finds its predominant application when genes that show similar expression patterns are clustered together. This kind of analysis may help, for example, to identify genes that share the same regulatory mechanisms or functions [1, 2].

M.C.P. de Souto and M.G. Kann (Eds.): BSB 2012, LNBI 7409, pp. 120–131, 2012.

A somewhat different kind of application is obtained when patient tissue samples are clustered together. In this type of analysis the main goal is to detect previously unknown clusters of tissue samples, which are usually associated with unknown types or subtypes of cancer. Since the work presented by Golub et al. [3], clustering of sample tissues has drawn quite an attention of the research community and has been employed in a substantial number of works, e.g., [4–7].

Tissue clustering problems are characterized by their small number of samples which are, in turn, embedded in a high dimensional feature space composed by thousands of genes. In order to cope with this particular scenario, a great variety of clustering algorithms has been both employed and developed. Despite of the great diversity of clustering algorithms, not until recently some guidelines have been established concerning on how to choose a specific method for clustering cancer samples. Comparative studies regarding different algorithms are presented in [8–10], whereas theoretical reviews of clustering algorithms considering their application to gene expression data are presented in [11, 2]. We note, that albeit important, the choice of the clustering algorithm itself is not the only determinant factor of clustering results. In fact, the choice of an appropriate proximity measure (similarity or distance) employed between pairs of objects is often regarded as a central issue in cluster analysis, since different measures may produce quite different results [12–15]. Even playing a key role in clustering, little attention has been given to the particular subject of proximity measures in gene expression [13–15]. Considering the clustering of tissue samples, this becomes a major issue, since the most commonly employed clustering algorithms are based on the definition of a proximity measure [16, 2, 17].

In order to give some idea about the importance of choosing an appropriate proximity measure when performing clustering analysis, we recall that the comparison of different proximity measures has been carried out in diverse areas of expertise, such as text clustering [18] and clustering-based feature selection [19], just to mention a few. It is important to note that a particular type of proximity measure is usually preferred given the characteristics of the problem at hand. For gene expression, it is accepted that objects should be regarded as similar if they exhibit similarity in shape (trend similarity), rather than in the absolute differences in their values [20]. As a consequence, the Pearson correlation coefficient has been widely adopted as the rule of thumb for clustering gene expression data [1, 15, 21], even though it is not the only correlation coefficient[1] available.

Regarding the tissue clustering scenario, some authors considered different proximity measures in their studies. In [9, 10], for instance, the authors reckon the importance of considering different proximity measures during tissue sample clustering. In these two studies, however, the authors are primarily interested in the comparison of clustering algorithms rather than the proximity measures themselves. In [22, 16, 23] the main focus of the authors is indeed the comparison of different proximity measures. The primary focus of [22] is to show proximity measures that may be considered as alternatives regarding the gene clustering scenario. Besides being a different scenario than the one addressed

---

[1] We use the terms correlation coefficient and correlation interchangeably in this paper.

here, we note that no guidelines is presented to the practitioner. In [16, 23] proximity measures are compared to both tissue and gene clustering. Although closely related at a first glance, tissue and gene clustering problems usually face different difficulties. Whereas in tissue sample clustering one has few high dimensional objects (samples), in gene clustering there is a large number of very low dimensional objects (genes). To this extent, conclusions from [16, 23] may be biased and misleading. Finally, we note that none of the previous studies focused on the evaluation of different types of correlation coefficients as alternatives to the commonly employed Pearson, even though this particular kind of measure is capable of capturing the similarity usually sought in gene expression data.

Bearing such considerations in mind, we present, to the best of our knowledge, the first comparison specifically designed to evaluate different correlation coefficients for clustering cancer samples. We advance the previously discussed works, at least in the following directions. Our primarily focus is the comparison of correlation coefficients, as this type of measure is amongst the most popular ones for gene expression data. We compare five correlation coefficients (along with Euclidean distance, which is also commonly employed in gene expression clustering), regarding: (i) their intrinsic separation ability and (ii) their predictive ability regarding four clustering algorithms. Differently from previous studies, however, we consider only the tissue clustering scenario, providing results regarding the two technologies from which gene expression data is usually available, i.e., Affymetrix and cDNA [9]. In such a manner, we provide not only detailed information regarding the choice of a particular proximity measure, but also a more fair comparison than the ones performed by previous works.

The remainder of the paper is organized as follows. In Section 2, we briefly review the five correlation coefficients considered in our study. In Section 3 we provide detail about both methodology and experimental setup. The main results of our work are presented in Section 4. Finally, in Section 5 we draw the conclusion and elaborate on future work regarding proximity measure evaluation.

2   Correlation Coefficients

Considering gene expression data, two samples are usually regarded as similar if they exhibit similarity in shape (trend), rather than in absolute differences from their values. Bearing that in mind, correlation coefficients have been widely used, as they capture such kind of similarity. Any two samples can be seen as real valued sequences $\mathbf{a} = (a_1, \ldots, a_n)$ and $\mathbf{b} = (b_1, \ldots, b_n)$, for which a correlation coefficient can be directly applied. During our evaluation we adapted each correlation as a dissimilarity measure between tissue samples in the form[2]:

$$\text{distance}(\mathbf{a}, \mathbf{b}) = 1 - \text{correlation coefficient}(\mathbf{a}, \mathbf{b}).$$

In the sequel we review the correlation coefficients considered in our study.

---

[2] Note that other adaptations may also be taken into account, although the aforementioned one is (as far as we know) the most employed one [9, 16, 22–24].

2.1   Pearson

The Pearson correlation coefficient (PE) [25] allows the identification of linear correlations between sequences. It is given by Eq. (1), where $\bar{a}$ and $\bar{b}$ are the means of $\mathbf{a}$ and $\mathbf{b}$. Due to its sensitivity to outliers, Pearson may produce false positives, i.e., sequence pairs that are not alike but receive a high correlation value. It has $O(n)$ running time.

$$PE(\mathbf{a}, \mathbf{b}) = \frac{\sum_{i=1}^{n}(a_i - \bar{a})(b_i - \bar{b})}{\sqrt{\sum_{i=1}^{n}(a_i - \bar{a})^2}\sqrt{\sum_{i=1}^{n}(b_i - \bar{b})^2}} \tag{1}$$

2.2   Jackknife

The underlying idea of the Jackknife correlation (JK) [20] is to minimize the effect of single outliers in the final correlation value. This is accomplished by removing one single value at a time from both sequences. If the sequences do not contain outliers, their correlation value will remain stable. In case the sequences have outliers, their removal will cause a decrease in their correlation, indicating that the sequences were correlated partly due to the presence of outliers. Jackknife is defined in Eq. (2), where $PE^i(\mathbf{a}, \mathbf{b})$ is the Pearson correlation between sequences $\mathbf{a}$ and $\mathbf{b}$ with their i^{th} values removed. Its time complexity is $O(n^2)$.

$$JK(\mathbf{a}, \mathbf{b}) = min\{PE^1(\mathbf{a}, \mathbf{b}), \ldots, PE^n(\mathbf{a}, \mathbf{b}), PE(\mathbf{a}, \mathbf{b})\} \tag{2}$$

2.3   Spearman

The Spearman correlation (SP) [26] can be seen as a particular case of the Pearson correlation, provided that values of both $\mathbf{a}$ and $\mathbf{b}$ are replaced with their ranks in the respective sequences. Bearing such in mind, Spearman can also be defined by Eq. (1). As only the ranks of the sequences are considered, Spearman is more robust to outliers than Pearson. Spearman has also been employed in the analysis of gene expression data [2]. Its has a running time of $O(n \ log \ n)$.

2.4   Kendall

The Kendall correlation (KE) [27] takes into account only the ranks of $\mathbf{a}$ and $\mathbf{b}$. It is defined according to the number of concordant (S_+), discordant (S_-), and neutral pairs in the sequences. In a concordant pair, the same relative order applies in the two sequences, i.e., $a_i < a_j$ and $b_i < b_j$ or $a_i > a_j$ and $b_i > b_j$. For discordant pairs, the inverse relative order applies, i.e., $a_i < a_j$ and $b_i > b_j$ or $a_i > a_j$ and $b_i < b_j$. All other pairs are deemed as neutrals. Kendall is defined in Eq. (3), where $n(n - 1)/2$ is the number of pairs. Extreme values are obtained only in the absence of neutrals. It has $O(n \ log \ n)$ time complexity.

$$KE(\mathbf{a}, \mathbf{b}) = \frac{S_+ - S_-}{n(n - 1)/2} \tag{3}$$

2.5 Rank-Magnitude

The Rank-Magnitude correlation (RM) [28] was introduced as an asymmetric measure, for cases in which one of the sequences is composed by ranks and the other by real values. It is defined by Eq. (4), with $R(a_i)$ denoting the rank of the ith position for sequence $\mathbf{a}$. In Eq. (4), $r^{min} = \sum_{i=1}^{n}(n + 1 - i)\bar{b}_i$ and $r^{max} = \sum_{i=1}^{n} i\bar{b}_i$. Value $\bar{b}_i$ corresponds to the ith element of the sequence, which is obtained by rearranging sequence $\mathbf{b}$ so that it gets sorted in ascending order.

$$\hat{r}(\mathbf{a}, \mathbf{b}) = \frac{2 \sum_{i=1}^{p} R(a_i)b_i - r^{max} - r^{min}}{r^{max} - r^{min}} \qquad (4)$$

We employed a symmetric version of Rank-Magnitude, given by $RM(\mathbf{a}, \mathbf{b}) = (\hat{r}(\mathbf{a}, \mathbf{b}) + \hat{r}(\mathbf{b}, \mathbf{a}))/2$. Any mention to Rank-Magnitude in this paper refers to its symmetric version, whose time complexity is $O(n \, log \, n)$.

3 Evaluating Proximity Measures

We take into account five different correlation coefficients in our comparison, namely, Pearson (PE), Jackknife (JK), Kendall (KE), Spearman (SP) and Rank-Magnitude (RM). In order to provide a fair comparison among these measures, we also include in our comparison the well-known Euclidean distance (EUC), which is also commonly employed to the clustering of cancer samples [9, 16, 1].

During the evaluation of the intrinsic separation and the predictive ability of the measures we consider the 35 labeled benchmark datasets proposed by [9]. In brief, this publicly available benchmark collection encompasses 35 microarray datasets from cancer gene expression experiments and comprehend the two platforms in which the technology is generally available, i.e., Affymetrix (21 datasets) and cDNA (14 datasets). All datasets are already preprocessed, with the most significant preprocessing performed by [9] being related to the removal of uninformative genes (genes that are not differentially expressed across samples). Information concerning datasets and preprocessing may be obtained in [9]. In the sequel we provide further detail on our evaluation methodology.

3.1 Intrinsic Separation Ability

The intrinsic separation capability of a proximity measure (in our case a correlation) indicates how well it is able to separate data points without the influence of a clustering algorithm [16]. To measure the separation ability of each one of the correlations we adopt the same methodology employed in [16]. Given a dataset with m samples, a distance matrix D, where $D(i, j) = \text{distance}(\mathbf{x_i}, \mathbf{x_j})$, with $1 \leq i, j \leq m$, may be easily constructed. Assuming that all the values of D are in the interval $[0, 1]$, we may proceed and build a binary classifier that assigns data points to each one of the two classes according to Eq. (5), where $i = ((l - 1) \; div \; m)$, $j = ((l - 1) \; mod \; m)$ and $\phi \in [0, 1]$ is a given threshold.

$$I_\phi(l) = \begin{cases} 1 & \text{if } D(i,j) \leq \phi \\ 0 & \text{otherwise} \end{cases} \tag{5}$$

Once we are dealing with labeled data we may build a desired solution for the classifier previously described. The desired solution is build upon the golden standard partition of each dataset as given by Eq. (6), for all pairs $\mathbf{x_i}$ and $\mathbf{x_j}$.

$$J(l) = \begin{cases} 1 & \text{if } \mathbf{x_i} \text{ and } \mathbf{x_j} \text{ belong} \\ & \text{to the same cluster} \\ 0 & \text{otherwise} \end{cases} \tag{6}$$

Note that by considering values of ϕ in the interval $[0, 1]$ in Eq. (5) one may evaluate the separation capability of a distance (in this case, a correlation coefficient) against an expected solution, as given by Eq. (6). In brief, these multiple comparisons (one per each value of ϕ) may be addressed by Receiver Operating Characteristics analysis, ROC analysis for short [29, 16, 23]. We compare the intrinsic separation ability of each distance by its Area Under the Curve (AUC).

3.2   Predictive Clustering Ability

The intrinsic separation ability of a distance (in our case, a correlation coefficient) may provide valuable information about its performance for a particular type of data. We recall, however, that we are primarily interested in the performance of different correlation coefficients regarding the tissue clustering problem.

To provide information about how different correlations perform with commonly employed clustering algorithms we generate partitions containing the same number of clusters as defined by the golden standard partition, i.e., the original labeling of each dataset. Result partitions are then compared based on their Adjusted Rand values [30], which evaluate the capability of each correlation in recovering partitions that are in conformity with the a priori structure defined in the data. We choose to employ the Adjusted Rand due to its correction that takes into account conformities between partitions found by chance [30, 12].

To evaluate the predictive ability of the correlation coefficients we consider four clustering algorithms commonly employed to tissue clustering problems [2, 9–11]: Single-Linkage (SL), Average-Linkage (AL), Complete-Linkage (CL) and k-medoids (KM). At this point, it is important to explain our preference for KM over the more popular k-means. Considering k-means and the well-known Euclidean distance, the centroid of each cluster is defined by the arithmetic mean of the objects that belong to it. Note, however, that for other proximity measures the centroid calculation must be redefined in order to maintain k-means optimization and convergence proprieties [31]. To avoid convergence problems, we choose KM, a counterpart of k-means in which the centroid (mean vector) is replaced by the medoid (most representative object in the cluster). Still regarding KM, we note that for each dataset and different correlation coefficient it is executed 30 times (random starts) as it is not deterministic. At the end of the 30 executions, we select only the best result partition for further comparison.

Finally, it is important to make clear that we only compare *different* correlation coefficients when considering the *same* clustering algorithm, as different clustering algorithms may provide different biases, e.g., we can compare PE and KE considering KM, but we *cannot* compare PE considering KM and KE considering CL, for instance. Note that in the former example, we are indeed comparing the different correlation coefficients, as the clustering algorithm employed is the same to both measures. In the latter example, however, we are comparing the pairs that encompass the clustering algorithm and the correlation coefficient.

3.3   Statistical Significance

In order to provide reassurance about the validity of our results, we employ the Friedman and Nemenyi statistical tests (with a 95% confidence level), which are more appropriate when comparing multiple algorithms on multiple datasets [32]. Regarding the intrinsic separation ability scenario, the application of the statistical tests is straightforward. For the predictive clustering ability scenario, we employ the tests considering each clustering algorithm separately, i.e., we apply the tests to compare the correlations separately within clustering algorithms.

4   Results and Discussion

4.1   Intrinsic Separation Ability

We summarize in Figure 1 results regarding the intrinsic separation ability of the correlations by their mean AUC results. For cDNA, PE and JK provided the best results, followed by RM. Regarding Affymetrix datasets, RM and JK displayed the best results, closely followed by PE. Regardless of the type of data, rank-based measures (KE and SP) displayed worse results than the other correlation coefficients. This behavior may be observed as a consequence of the loss of information that is intrinsic to rank-based measures. Finally, EUC presented the worst overall AUC results for both cDNA and Affymetrix data.

To provide reassurance about the validity of the AUC results we employed Friedman and Nemenyi statistical tests (at a 95% confidence level), separately for cDNA and Affymetrix data. For cDNA data the tests suggest that Pearson, Jackknife and Rank-Magnitude provided better intrinsic separation ability than Euclidean distance. Considering Affymetrix data, the tests suggest that Rank-Magnitude presented superior intrinsic separability than Euclidean distance.

4.2   Predictive Clustering Ability

We summarize the results of the predictive clustering ability in Figure 2. We depict mean Adjusted Rand values for the 14 cDNA and 21 Affymetrix datasets.

Considering AL and Affymetrix data, practically all the correlation coefficients displayed similar mean Adjusted Rand results. For cDNA datasets, JK and RM presented the best mean results. Still regarding this type of data the commonly

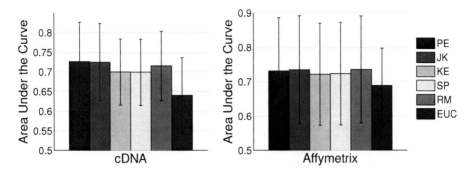

Fig. 1. Intrinsic separation ability of each one of the evaluated distances. Figure depicts the mean Area Under the Curve (AUC) values. Bars display mean results for each correlation in different types of datasets (error bars account for standard deviation).

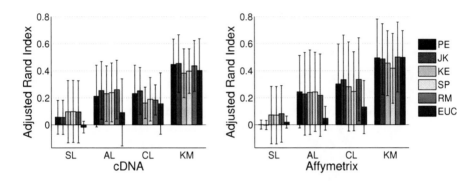

Fig. 2. Class recovery considering the same number of clusters as defined by the golden standard partitions. Bars display mean results obtained for pairs of clustering algorithm and distance in different types of datasets (error bars account for standard deviation).

employed PE provided the worst results among the correlation coefficients. Note that for both cDNA and Affymetrix datasets, rank-based measures, i.e., KE and SP, presented similar behavior among themselves, whereas the commonly employed Euclidean distance showed the worst overall results.

For CL, when considering cDNA data, both PE and JK provided the best results. For Affymetrix data, particularly, RM showed along with JK the best overall results. Close results were found when comparing rank-based correlation coefficients (KE and SP). The commonly employed Euclidean distance (EUC) provided, once more, the worst overall results regardless of the data type.

Regarding the k-medoids clustering algorithm (KM) three measures figured out as the top ones, regardless of the type of data under analysis, namely: RM, JK and PE. It is interesting to note that for KM the commonly employed Euclidean distance (EUC) provided competitive results with the top ranked correlation coefficients, mainly for Affymetrix datasets. Finally, once more, rank-based correlation (KE and SP) showed poorer results than other correlations.

SL led to the poorest recovery rates among the clustering algorithms employed. Note, however, that for this particular algorithm, correlation coefficients that take into account the magnitude of each sample (PE and JK), displayed the worst mean results among the evaluated correlation coefficients. A possible explanation for such a behavior may be the fact that SL is very sensitive to the presence of outliers and noise. This sensibility may be pronounced with the use of these two specific correlation coefficients and alleviated with the adoption of rank-based measures, such as KE and SP. To this extent, our results reinforce the results presented in [9], because even with the use of different correlations, SL clearly does not stand as good choice for the tissue clustering scenario.

Finally, we applied Friedman and Nemenyi statistical tests (at a 95% confidence level), separately for each algorithm, to detect if one or more distances provide statistically significant differences regarding their Adjusted Rand values. Although the Friedman statistical test suggests that there were differences among the correlation coefficients under comparison the Nemenyi statistical test was unable to identify between which pairs these differences were found.

4.3   Relation between Intrinsic Separation and Predictive Ability

As pointed out by [16, 23], one may proceed and compare the intrinsic separation ability of a particular distance with its predictive ability, evaluating the interplay of both. Differently from [23], we do not compare the performances of the distances themselves with their performance in conjunction with different clustering algorithms. We show, however, that these two analysis are indeed related and consistent by correlating the values of Area Under the Curve and Adjusted Rand. In such a way, one may expect high correlation values if the two analysis are consistent and low correlation values when dealing with inconsistent results.

To this extent, both analysis performed here turn out to be quite consistent when compared against each other. For instance, when correlating Area Under the Curve and Adjusted Rand values we found a mean Spearman correlation coefficient of 0.82 for the experiments with KM and 0.74 for the experiments with CL. The only exception is found when correlating Area Under the Curve and Adjusted Rand values for the SL algorithm, which provides the worst results.

4.4   Discussion

First of all, it is important to note that the results here presented show a high variability, i.e., quite different results are obtained for different datasets, as also noted in [9, 10]. As a consequence, on some datasets, a particular correlation that does not provide the best mean results, may turn out to be the best choice. Considering results in general, it is fair to say that some trends were observed:

- All correlation coefficients provided better results than EUC, except for KM. Note, however, that for this particular clustering algorithm, PE, JK and RM provided competitive or better results than EUC. This behavior is in agreement with the fact that samples should be regarded as similar if they

exhibit similarity in shape (trend), rather than in absolute differences from their values. Based on such results, regardless of the clustering algorithm one should prefer employing a correlation rather than Euclidean distance.
- PE and JK (which is based on PE) displayed superior results when compared to other correlations. Although JK has a higher computational cost than PE, it may provide better results than PE in the presence of outliers. If one is willing to pay JK quadratic time complexity, it should be preferred over PE.
- RM, which has an intermediate computational complexity between PE and JK, provides, in a number of cases, competitive results to both. Whereas PE and JK are based solely on the magnitude values of the sequences, RM considers also their ranks. Thus, RM may be more robust to noise than PE and JK and arises as a good alternative to both correlation coefficients. In fact, for noisy data, we argue that RM is preferable over both PE and JK.
- Rank-based correlations showed worst results than other correlation coefficients, such as PE, JK and RM. This particular behavior may be explained by the loss of information inherent in the definition of these measures. However, when dealing with noisy data, rank-based measures may turn out to be valuable choices. To this extent, KE showed in a number of cases better results than the commonly employed SP, providing good alternatives to it.

It is important to note that none of the correlations considered in our study outperformed the others in all datasets. This fact may be associated to the differences observed both in the datasets and in the formulation of the correlation coefficients themselves. To this extent, when considering a specific problem, it may be interesting to perform a preliminary analysis of different measures and, then, choose the one that produces the best results. If no correlation dominates the others, then the less computationally expensive one may be adopted.

5   Conclusions and Future Work

We have presented a comparison of five correlation coefficients (along with Euclidean distance) for clustering cancer samples of gene expression microarray data. In a first step, we evaluated the intrinsic separation ability of the correlation coefficients. Further on, we evaluated their predictive performance when employed with different clustering algorithms. To this extent, we employed four clustering algorithms frequently used in clustering of tissue samples problems.

As future work we intend to include other correlation coefficients and even "classical" proximity measures to our comparison, such as: Manhattan, Supreme and Cosine. Regarding the predictive clustering ability of the measures, we intend to evaluate other scenarios, considering different number of clusters than the ones defined by the reference partitions, simulating thus real applications in which the user do not have a initial clue about the number of clusters present in the data. We also intend to compare different proximity measures regarding the gene clustering scenario. To this extent, we aim to consider proximity measures that were specifically developed to the gene clustering scenario, i.e., [20, 33–35], as to this date none of these measures has been systematically evaluated. Finally, we pretend to evaluate the behavior of the proximity measures in the presence of different levels of noise, assessing their effect in the performance of the measures.

Acknowledgements. The authors would like to acknowledge Brazilian research agencies CAPES, CNPq, FACEPE and FAPESP (process #2011/04247-5) for financial support.

References

1. D'haeseleer, P.: How does gene expression clustering work? Nature Biotechnology 23, 1499–1501 (2005)
2. Kerr, G., Ruskin, H.J., Crane, M., Doolan, P.: Techniques for clustering gene expression data. Computers in Biology and Medicine 38(3), 283–293 (2008)
3. Golub, T.R., et al.: Molecular classification of cancer: Class discovery and class prediction by gene expression monitoring. Science 286(5439), 531–537 (1999)
4. Alon, U., et al.: Broad patterns of gene expression revealed by clustering analysis of tumor and normal colon tissues probed by oligonucleotide arrays. Proceedings of the National Academy of Sciences 96(12), 6745–6750 (1999)
5. Alizadeh, A.A., et al.: Distinct types of diffuse large b-cell lymphoma identified by gene expression profiling. Nature 403(6769), 503–511 (2000)
6. Ramaswamy, S., Ross, K.N., Lander, E.S., Golub, T.R.: A molecular signature of metastasis in primary solid tumors. Nature Genetics 33(1), 49–54 (2003)
7. Lapointe, J., et al.: Gene expression profiling identifies clinically relevant subtypes of prostate cancer. Proceedings of the National Academy of Sciences 101(3), 811–816 (2004)
8. Pirooznia, M., Yang, J., Yang, M.Q., Deng, Y.: A comparative study of different machine learning methods on microarray gene expression data. BMC Genomics 9(suppl. 1), S13 (2008)
9. Souto, M., Costa, I., de Araujo, D., Ludermir, T., Schliep, A.: Clustering cancer gene expression data: A comparative study. BMC Bioinformatics 9(1), 497 (2008)
10. Freyhult, E., Landfors, M., Onskog, J., Hvidsten, T., Ryden, P.: Challenges in microarray class discovery: A comprehensive examination of normalization, gene selection and clustering. BMC Bioinformatics 11(1), 503 (2010)
11. Jiang, D., Tang, C., Zhang, A.: Cluster analysis for gene expression data: A survey. IEEE Transactions on Knowledge and Data Engineering 16(11), 1370–1386 (2004)
12. Jain, A.K., Dubes, R.C.: Algorithms for Clustering Data. Prentice-Hall, Inc., Upper Saddle River (1988)
13. Brazma, A., Vilo, J.: Gene expression data analysis. FEBS Letters 480(1), 17–24 (2000)
14. Steuer, R., Kurths, J., Daub, C.O., Weise, J., Selbig, J.: The mutual information: Detecting and evaluating dependencies between variables. Bioinformatics 18(suppl. 2), S231–S240 (2002)
15. Priness, I., Maimon, O., Ben-Gal, I.: Evaluation of gene-expression clustering via mutual information distance measure. BMC Bioinformatics 8(1), 111 (2007)
16. Giancarlo, R., Lo Bosco, G., Pinello, L.: Distance Functions, Clustering Algorithms and Microarray Data Analysis. In: Blum, C., Battiti, R. (eds.) LION 4. LNCS, vol. 6073, pp. 125–138. Springer, Heidelberg (2010)

17. Souto, M.C.P., de Araujo, D.S.A., Costa, I.G., Soares, R.G.F., Ludermir, T.B., Schliep, A.: Comparative study on normalization procedures for cluster analysis of gene expression datasets. In: IJCNN, Hong Kong, China, pp. 2792–2798. IEEE (2008)

18. Boyack, K.W., et al.: Clustering more than two million biomedical publications: Comparing the accuracies of nine text-based similarity approaches. PLoS ONE 6(3), e18029 (2011)

19. Jaskowiak, P.A., Campello, R.J.G.B., Covões, T.F., Hruschka, E.R.: A comparative study on the use of correlation coefficients for redundant feature elimination. In: 11th Brazilian Symposium on Neural Networks, São Paulo - Brazil, pp. 13–18 (2010)

20. Heyer, L.J., Kruglyak, S., Yooseph, S.: Exploring expression data: Identification and analysis of coexpressed genes. Genome Res. 9(11), 1106–1115 (1999)

21. Loganantharaj, R., Cheepala, S., Clifford, J.: Metric for measuring the effectiveness of clustering of DNA microarray expression. BMC Bioinformatics 7, S5 (2006)

22. Gentleman, R., Ding, B., Dudoit, S., Ibrahim, J.: Distance measures in DNA microarray data analysis. In: Bioinformatics and Computational Biology Solutions Using R and Bioconductor, pp. 189–208. Springer, New York (2005)

23. Giancarlo, R., Lo Bosco, G., Pinello, L., Utro, F.: The Three Steps of Clustering in the Post-Genomic Era: A Synopsis. In: Rizzo, R., Lisboa, P.J.G. (eds.) CIBB 2010. LNCS, vol. 6685, pp. 13–30. Springer, Heidelberg (2011)

24. Jaskowiak, P.A., Campello, R.J.G.B.: Comparing correlation coefficients as dissimilarity measures for cancer classification in gene expression data. In: 6th Brazilian Symposium on Bioinformatics, Brasília - Brazil, pp. 1–8 (2011)

25. Pearson, K.: Contributions to the mathematical theory of evolution. iii. Regression, heredity, and panmixia. P. Roy. Soc. Lond. A Mat. 59, 69–71 (1895)

26. Spearman, C.: The proof and measurement of association between two things. Am. J. Psychol. 100(3/4), 441–471 (1904)

27. Kendall, M.G.: Rank Correlation Methods, 4th edn. Griffin, London (1970)

28. Campello, R.J.G.B., Hruschka, E.R.: On comparing two sequences of numbers and its applications to clustering analysis. Inform. Sciences 179(8), 1025–1039 (2009)

29. Hand, D.J., Till, R.J.: A simple generalisation of the area under the ROC curve for multiple class classification problems. Machine Learning 45, 171–186 (2001)

30. Hubert, L., Arabie, P.: Comparing partitions. Journal of Classification 2, 193–218 (1985)

31. Steinley, D.: K-means clustering: A half-century synthesis. British Journal of Mathematical and Statistical Psychology 59, 1–34 (2006)

32. Demšar, J.: Statistical comparisons of classifiers over multiple data sets. J. Mach. Learn. Res. 7, 1–30 (2006)

33. Bolshakova, N., Azuaje, F.: Cluster validation techniques for genome expression data. Signal Processing 83(4), 825–833 (2003)

34. Möller-Levet, C.S., Klawonn, F., Cho, K.H., Yin, H., Wolkenhauer, O.: Clustering of unevenly sampled gene expression time-series data. Fuzzy Sets and Systems 152(1), 49–66 (2005)

35. Son, Y.S., Baek, J.: A modified correlation coefficient based similarity measure for clustering time-course gene expression data. Pattern Recognition Letters 29(3), 232–242 (2008)

Associating Genotype Sequence Properties to Haplotype Inference Errors

Rogério S. Rosa, Rafael H.S. Santos, and Katia S. Guimarães

Informatics Center,
Federal University of Pernambuco, UFPE
Recife, Brazil
{rsr,rhss,katiag}@cin.ufpe.br

Abstract. Haplotype analysis has become an important tool in studying species traits and susceptibility to diseases. Several computational methods for determining haplotype information from genotype data have been developed, but none is perfect. Haplotype Inference (HI) approaches based on different strategies or biological principles tend to fail in different loci. In this work we apply Multiple Linear Regression to explore the relevance of several biologically meaningful properties of the genotype sequences for the occurrence of errors in the results of three HI methods based on different principles. We develop models for databases on different elements, using two error metrics. We assess the accuracy of our results through statistical analysis. Our models reveal genotype properties that are relevant in general and others that are suited for particular scenarios. We also show that the Regression models present statistically better performance than Neural Network models developed for the same databases and properties.

Keywords: Genotype, Haplotype Inference, Linear Regression, Neural Network, Statistical Analysis.

1 Introduction

Haplotype information is valuable in understanding species evolution, as well as in association studies which try to correlate the propensity to certain diseases with patterns inherited through the haploid cells. Since capturing haplotypes directly from experiments is both difficult and expensive, it would be highly desirable to determine haplotypes through combinatorial or statistical approaches.

Inferring haplotype from genotypic data can be formally described as follows. A matrix $H_n m$ with n individuals and m single nucleotide polymorphisms (SNPs) presents $2n$ haplotypes. Each haplotype pair H_{2i-1} and H_{2i}, $1 \leq 1 \leq n$, generates genotype g with m SNPs. Each single position in g is called a *site* or *locus*. Genotype g can be computationally represented by a vector over alphabet 0. 1, 2, where symbols 2 represent heterozygous sites, and 0 and 1 represent homozygous sites. Then a genotype vector g, with m sites, can be explained (resolved) by two haplotype vectors h_1 and h_2, where each site $h_1(i)$ and $h_2(i)$, $1 \leq i \leq m$, has $h_1(i), h_2(i) \in \{0, 1\}$, and follows the rule given by Equation 1.

M.C.P. de Souto and M.G. Kann (Eds.): BSB 2012, LNBI 7409, pp. 132–143, 2012.

$$\begin{cases} h_1(i) = h_2(i) = g(i), \text{ if } g(i) \in \{0, 1\}; \\ h_1(i) = 1 - h_2(i), \quad \text{ if } g(i)=2. \end{cases} \tag{1}$$

Several methods have been proposed for HI, such as the Clark Method [2], Integer Programming formulations based on the parsimony principle [6] [8] [10] [9] [1], phylogeny based methods [7] [4], Bayesian methods [19] [15], and methods based on Markov Chain models [20] [22].Those approaches usually consider one of the main biological models: Parsimony or Perfect Phylogeny. In the perfect phylogeny model, the HI problem is linear [4], however, that approach requires that the sequences have not been subject to recombination events, which is hardly the case. HI by parsimony is an NP-hard Problem [6]. Approaches based on population statistics are usually more successful, but in general the error rates are very high, and the results frequently present bias. In a previous study, Rosa and Guimarães [16] showed that although different algorithms for HI may present similar error scores, most of the errors occur in different loci along the genotype sequence. That can be explained by the fact that the methods resort to different insights and strategies to approach the problem. It would not be feasible to combine several strategies in one single method, due to the high complexity of the problem and also because different approaches can even contradict each other occasionally.

Identifying regions of the genotype where each method has a higher propensity to make mistakes, could eventually lead to an ensemble approach, based on biological properties of the sequences. To determine which variables could be associated with inference errors of the different methods is a difficult problem. In a previous work [17], we investigated the power of neural network (NN) models to predict errors, and we assessed their performance using different error measures, in different types of data. In this work, we design multiple linear models to predict the inference error of three methods, Haplorec, fastPHASE, and PTG, and we statistically analyze their performances. Hypothesis tests are used to validate the results and show that the models are statistically significant and more accurate than the NN models.

The paper is organized as follows. In the next section the experiments and the data used are described. The results obtained are presented in Section 3. Section 4 contains a brief discussion of the results and the conclusions.

2   Experiments Design

2.1   Data Bases Used

The data set contains properties of sets of haplotypes /genotypes collected from project HapMap [3]. The genotype data is from Chromosome 20 of five different ethnic populations: Chinese in Denver (CHD), Indian in Houston (GIH), African Luhya in Kenia (LWK), African Maasai in Kenia (MKK), and Italians in Toscana (TSI). Each segment of genotype collected has 1000 SNPs. The Chromosome positions from where the sequences were taken were randomly selected.

Besides the properties of each population, the bases contain the error inference obtained by three haplotyping methods: fastPHASE [18], Haplorec [5], and PTG [11]. Each of these methods is based on a different computational technique. fastPHASE is a statistical method based on Maximum Likelihood, Haplorec uses Markov Models, while PTG is a combinatorial algorithm based on Parsimonious Tree-Grow. Those three approaches are widely used to solve large genotype datasets.

The errors of the Haplotype Inference, as well as the genotype properties were computed considering two distinct data bases: the Individuals Base and the SNPs Base. The properties collected from the genotype datasets were the following. (1) Number of symbols 0 ($NS0$); (2) Number of symbols 1 ($NS1$); (3) Heterozigocity (HTZ), given by the number of symbols 2; (4) Density of Heterozigocity (DHZ), represented by the number of neighbors with symbol 2 paired with symbols 2 in each element; (5) Number of blocks of symbols 2 ($NB2$), where a block is a sequence of identical symbols in an element; (6) Average length of blocks of symbols 2 ($LB2$); and (7) Conservation level (CSV), given by the Linkage Disequilibrium for SNPs, and by a Markov chain estimation for Individuals. For a fragment of genotype F with m SNPs of an individual, CSV is represented by probability p given by Equation 2 [5].

$$p(F) = p(F(1)) \prod_{i=2}^{m} p(F(i)|F(i-1)).$$ (2)

The Individuals and the SNPs bases contain the properties and the inference errors obtained by each method for each single element. The error metrics more frequently used for the HI Problem are Error Rate (ER) [15] and Switch Error (SE) [12]. For the Error Rate, the correct haplotypes are aligned to the inferred ones, the number of mismatches is computed, and the Error Rate is given by the ratio between the number of mismatches and the total number of sites in the dataset. The Swith Error is calculated as $(N - 1 - SD) / (N - 1)$, where N denotes the number of heterozigous loci, and SD (Switch Distance) denotes the number of exchanges required between the two inferred haplotypes, in order to make them identical to the original ones. Besides Error Rate, we used as a second error metric, SE for the Individuals Base, and SD for the SNPs Base.

The Individuals (SNPs) Base has 13 attributes: The 7 properties described, 3 results of Error Rates, and 3 results for Switch Error (Distance), one for each method considered. For each one of the six error attributes in the Individuals Base and the SNPs Base, a Multiple Linear Regression model was estimated and statistically validated, as described next.

2.2   Multiple Linear Regression Models

Twelve Multiple Linear Regression Models were estimated, six for the Individuals Base and six for the SNPs Base. Each model had as response variable one of the error attributes. The Step-Wise method [13] was applied to select among the seven explanatory variables (property attributes) a suitable model for each one of the response variables (error attributes). To the models suggested by Step-Wise, analysis of variance (ANOVA) for Regression [13] was applied, and explanatory variables with confidence level smaller than 95% were eliminated. The properties used in the models proposed by Step-Wise and validated by ANOVA are presented in Tables 1 and 2.

The training and test schemes were done with the following steps. (1) 15% of the samples for each population was randomly selected for testing, and the remaining samples were used for training; (2) For each answer variable, a regression model was estimated using the training dataset; (3) The models constructed in Step (2) were used to predict the errors in the test data; (4) Steps (1) through (3) were repeated 120 times, in order to generate a set of predicted samples large enough to execute t-tests with confidence greater than 95%; (5) Hypothesis tests were done with paired samples. All those tasks were done using the Statistics Software R [21].

2.3   Statistical Analysis of Results

The precision of the Haplotype Inference error was estimated for each one of the regression models. A set of hypothesis tests were designed to: (1) Assess the closeness between the values predicted by each regression model and the corresponding actual response values, and (2) Compare the performance of each regression model with the corresponding neural network model presented in [17]. In order to analyze the accuracy of the regression models estimated, we computed (1) The mean of the prediction errors and (2) Its standard deviation.

Three t-tests were performed to compare the actual values (A) and the predicted values (P), in order to collect evidence that the regression models estimated are relevant. The Null and Alternative Hypothesis were: (1) $H_0 : A \geq P$, $H_a : A < P$; (2) $H_0 : A \leq P$, $H_a : A > P$; and (3) $H_0 : A = P$, $H_a : A \neq P$. Ideally, none of the null hypothesis would be rejected. The results of the t-tests are presented in Tables 3 and 4.

We also compared the prediction errors of our Regression models to the errors of corresponding Neural Network models [17]. The error measure used in this case was not the raw error, but rather a relative error, a positive value given by $(|x - y|/x)100$, where x represents the actual value and y represents the predicted value. The relative error in each Regression model (E_R) and Neural Network model (E_{NN}) were computed.

The following hypothesis tests were applied: (1) $H_0 : E_R \geq E_{NN}$ and $H_a : E_R < E_{NN}$; and (2) $H_0 : E_R \leq E_{NN}$ and $H_a : E_r > E_{NN}$. The results obtained are presented in Table 5, for the Individuals Base, and in Table 7, for the SNPs Base.

3   Results

3.1   Selected Regression Variables

Individuals Base: Our results show that for the Individuals Base, the more significant models for response variable ER chosen by Step-Wise and by ANOVA were always different, with Step-Wise always taking one extra variable. For response variable SE, Step-Wise and ANOVA chose exactly the same models. Table 1 shows the models designed for each response variable. The best model for each response variable was the one indicated by the ANOVA test, because we wish to exclude properties not significant in the model. The regression variables in bold in Table 1 are those which presented 95% of confidence in the ANOVA test; the other variables presented 99% of confidence in that test.

Table 1. Regression models estimated by Step-Wise and more significant models as identified by ANOVA (with at least 95% confidence) for the Individuals Base

Error Type	Method	Step-Wise	ANOVA
ER	fastPHASE	$CSV + LB2 + NS0 + NS1$	$\mathbf{CSV} + NS0 + NS1$
	Haplorec	$CSV + NS0 + NS1$	$NS0 + NS1$
	PTG	$CSV + NB2 + NS0 + NS1$	$CSV + NS0 + NS1$
SE	fastPHASE	$CSV + NB2 + NS0 + NS1$	$CSV + \mathbf{NB2} + NS0 + NS1$
	Haplorec	$CSV + NB2 + NS0 + NS1$	$CSV + NB2 + NS0 + NS1$
	PTG	$CSV + DHZ + NS0$	$CSV + \mathbf{DHZ} + NS0$

SNPs Base: Table 2 shows the regression variables for the models estimated by the method Step-Wise, as well as the models with high significance according to ANOVA, for the SNPs Base. According to the ANOVA test, five of the six models estimated presented high significance for all variables of the Step-Wise models. The only exception was the model estimated for ER-Haplorec, for which regression variable DHZ was not significant, so it was discarded from the model. Variables HTZ, $NB2$, and $LB2$ presented 95% significance for models ER-PTG, SD-Haplorec, and SD-Haplorec and SD-PTG, respectively.

The variables selected by methods Step-Wise and ANOVA are exactly the same. Interestingly, regression variable CSV is significant for all models involving response variable ER, but is not significant for any model involving response variable SD. Also, regression variable $NB2$ is highly significant for all models.

3.2   Regression Models Accuracy

After identifying the statistically more significant models, the precision of each of those models was tested. A set of hypothesis tests were used to find evidence that the predicted error is on average the same as the actual error of each HI method.

Individuals Base: As shown in Table 3, none of the models estimated for the Individuals Base presented p-value < 0.05, for any of the three hypothesis tests considered, as described in Section 2.3. Of particular interest is the test for whether or not the predicted values are on average close to the actual values. Since the null hypothesis cannot be rejected, and the p-values are high, there is evidence that the actual values are statistically close to the ones predicted. The least relative error reported (|Actual - Predicted|) was in the model estimated for ER-PTG (0.00387), and the largest error was in model ER-Haplorec (0.02377).

Table 2. Regression models estimated by Step-Wise and more significant models as identified by ANOVA (with at least 95% confidence) for the SNPs Base

Error Type	Method	Step-Wise	ANOVA
ER	fastPHASE	$CSV+NB2+LB2+DHZ+$ $NS0+NS1+HTZ$	$CSV+NB2+LB2+DHZ+$ $NS0+NS1+HTZ$
	Haplorec	$CSV+NB2+LB2+DHZ+$ $NS0+NS1+HTZ$	$CSV+NB2+LB2+NS0+$ $NS1+HTZ$
	PTG	$CSV+NB2+LB2+DHZ+$ $NS0+NS1+HTZ$	$CSV+NB2+LB2+DHZ+$ $NS0+NS1+\textbf{HTZ}$
SD	fastPHASE	$NB2+DHZ+HTZ$	$NB2+DHZ+HTZ$
	Haplorec	$NB2+LB2+DHZ+NS0+$ HTZ	$\textbf{NB2+LB2}+DHZ+NS0+$ HTZ
	PTG	$NB2+LB2+DHZ+HTZ$	$NB2+\textbf{LB2}+DHZ+HTZ$

Table 3. Individuals Base: Errors obtained for each model and p-values of the hypothesis tests (with more than 95% of confidence), when comparing the actual values (A) with the predicted ones (P)

Error Type	Method	Error	σ	$H_a : A < P$	$H_a : A > P$	$H_a : A \neq P$
ER	fastPHASE	0.00804	0.00687	0.6981	0.3019	0.4248
	Haplorec	0.02377	0.01665	0.6602	0.3398	0.4465
	PTG	0.00387	0.00409	0.5483	0.4517	0.5192
SE	fastPHASE	0.01002	0.00821	0.5003	0.4997	0.5278
	Haplorec	0.01347	0.01145	0.4543	0.5457	0.5054
	PTG	0.02292	0.01747	0.4603	0.5396	0.4773

SNPs Base: Table 4 shows that there is evidence that the predicted values are on average equal to the actual error for five of the models, namely those with p-values > 0.05 for the three tests. The third hypothesis test ($H_a : A \neq P$) for model ER-fastPHASE presented p-value smaller than 0.05. On the other hand, the estimated model for ER-fastPHASE presented the second least error among all (0.00862). That could be explained by a low variance in both the actual and the predicted errors. From Table 4, one can see that the latter is indeed very low ($\sigma = 0.00735$).

3.3   Comparison of Regression and Neural Net Models

In a previous work [17], Neural Network (NN) models were constructed to predict the response variables from single or double input based on the same property variables. The errors in those predictions were statistically assessed, and the variables that presented the higher accuracy were identified. In this section we compare the accuracy of the two approaches, based on the results obtained from the same input data for the same response variables.

For the t-tests, the relative error is considered, that is the distance between the actual data and the data predicted by the regression models data are larger or smaller than the ones predicted by the best corresponding NN models.

Table 4. SNPs Base: Errors obtained for each model and p-values of the hypothesis tests (with more than 95% of confidence), when comparing the actual values (A) with the predicted ones (P)

Error Type	Method	Error	σ	$H_a : A < P$	$H_a : A > P$	$H_a : A \neq P$
	fastPHASE	0.00862	0.00735	0.9881	0.0119	0.0238
ER	Haplorec	0.00831	0.00945	0.5079	0.4921	0.5007
	PTG	0.01272	0.01036	0.5682	0.4318	0.4841
	fastPHASE	1.32625	2.21294	0.6030	0.3969	0.3952
SD	Haplorec	2.02898	3.09159	0.3533	0.6467	0.4397
	PTG	1.52855	2.46303	0.5457	0.4544	0.4153

Individuals Base: An interesting initial observation is that for response variable ER, the property variables used in the best performing regression models were almost always different from those that led to the best NN models. Table 5 presents the property variables that led to the best models for each algorithm, along with the p-values of the Null Hypothesis $E_R > E_{NN}$ ($E_R < E_{NN}$). The p-values indicate that there is strong evidence that the average of the relative error of each regression model is statistically smaller than the relative error of the corresponding NN model.

Properties CSV, $NS0$, and $NS1$ were selected for the Regression models of response variable ER for all three algorithms, while in the NN models, CSV was selected only for the PTG Algorithm, $NS0$ was chosen only for the Haplorec Algorithm, and $NS1$ was not used. On the other hand, property HTZ was used an input variable in the best NN models for two algorithms, while it was not selected in any of the regression models. Some properties were used in both models for response variable ER: Property $LB2$ for Algorithm fastPHASE, property $NS0$ for Algorithm Haplorec, and property CSV algorithm PTG. It can also be seen in Table 5 that property CSV was used in the two models for response variable SE for the three algorithms. Variable $NS0$ was selected in all models, except for the NN model for Algorithm PTG.

Table 6 presents the mean error, the mean deviation, and the maximum relative error found for each model for the Individuals Base. Considering the mean error, Regression lead to better models than NN in general. The best models were the ones constructed for response variable ER-fastPHASE, with errors 1.045% and 1.359achieved the best maximum errors, 5.115% for the Regression models, and 6.187%, for the NN model. In all cases, the relative error of the Regression models were smaller than the errors of the corresponding NN models. In the models for the ER response variable, the difference is highly evident. The Regression and the NN models for response variable ER-Haplorec both had a very poor performance.

For response variables (A) ER-fastPHASE, (B) ER-Haplorec, and (C) ER-PTG, Figure 1 plots, for a given x, the number of test cases with relative error value less than $x\%$. As the relative error grows (X axis), the number of cases also grows, since it is cumulative; of course, curves closer to the Y axis with a short tail represent more accurate models. The Y axis itself would represent a perfect prediction curve (error of 0%).

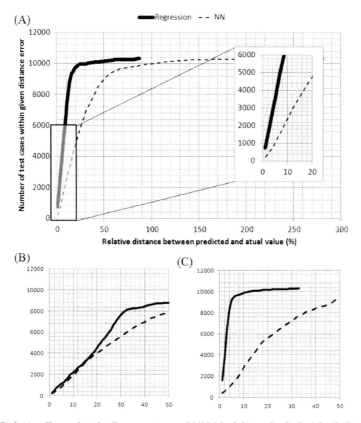

Fig. 1. Relative Error for the Regression and NN Models in the Individuals Base, for response variables (A) ER-fastPHASE,(B) ER-Haplorec, and (C) ER-PTG. In the X axis, the relative error, and in the Y axis, the number of test cases with relative error up to $X\%$.

Table 5. Individuals Base: p-values of the Comparison Between Regression and Neural Network Models

Error Type	Method	Regression Model	NN Model	$H_a : E_R < E_{NN}$	$H_a : E_R > E_{NN}$
ER	fastPHASE	$CSV + LB2 + NS0 + NS1$	$HTZ + LB2$	0	1
	Haplorec	$CSV + NS0 + NS1$	$DHZ + NS0$	0.0572	0.9428
	PTG	$CSV + NB2 + NS0 + NS1$	$CSV + HTZ$	0	1
SE	fastPHASE	$CSV + NB2 + NS0 + NS1$	$CSV + NS0$	1.71e-297	1
	Haplorec	$CSV + NB2 + NS0 + NS1$	$CSV + NS0$	1.13e-237	1
	PTG	$CSV + DHZ + NS0$	$CSV + HTZ$	5.77e-123	1

Figures 1(A) and 1(C) show that the Regression models are much more accurate than the NN models for response variables ER-fastPHASE and ER-PTG. The two models are much less accurate for response variable ER-Haplorec, nonetheless, the regression model is still slightly better.

SNPs Base: In the SNPs base, the regression models consistently used more variables than in the Individuals Base, and for two response variables, ER-fastPHASE and ER-PTG, all 7 properties were chosen. Ironically, three of the best NN models were obtained using only one property, including the two for which the regression model selected 7 inputs. For the NN models, variable HTZ, used in 5 of the 6 best models, was clearly the most informative property.

Table 6. Individuals Base: Relative Errors (%) in Regression and Neural Network Models

Error Type	Method	Regression Models			Neural Network Models		
		Mean	σ	Max	Mean	σ	Max
ER	fastPHASE	9.171	9.977	86.080	28.580	27.713	281.434
	Haplorec	62.111	507.480	11377.810	64.852	346.734	7686.342
	PTG	3.316	3.926	32.445	25.045	24.583	217.698
SE/SD	fastPHASE	1.045	0.867	5.115	1.359	1.025	6.187
	Haplorec	1.420	1.230	7.267	1.776	1.433	8.613
	PTG	4.164	2.960	17.039	4.813	3.396	19.074

Table 7. SNPs Base: p-values of the Comparisons between Regression E_R and Neural Network Models E_{NN}.

Error Type	Method	Regression Model	NN Model	$H_a : E_R < E_{NN}$	$H_a : E_R > E_{NN}$
ER	fastPHASE	$CSV+NB2+LB2+DHZ+NS0+NS1+HTZ$	HTZ	0	1
	Haplorec	$CSV+NB2+LB2+NS0+NS1+HTZ$	$NB2+NS0$	0	1
	PTG	$CSV+NB2+LB2+DHZ+NS0+NS1+HTZ$	HTZ	0	1
SD	fastPHASE	$NB2 + DHZ+HTZ$	$DHZ+HTZ$	9.15e-166	1
	Haplorec	$NB2+LB2+DHZ+NS0+HTZ$	HTZ	0	1
	PTG	$NB2+LB2+DHZ+HTZ$	$DHZ+HTZ$	9.15e-166	1

Once again, the p-values of the hypothesis tests, shown in Table 7, show that there is strong evidence that the average errors of the predictions by the Regression models are smaller than the average error prediction by the corresponding NN models, which is confirmed by the relative errors presented in Table 8, since the mean errors for the Regression models are smaller than the mean errors of the corresponding NN models. In fact, all models for the SNPs base had poor performance, except for the ones for response variable ER-PTG, suggesting that the properties chosen are not suitable for SNP elements.

Table 8. SNPs Base: Relative Errors (%) in Regression and Neural Network Models

Error Type	Method	Regression Models			Neural Network Models		
		Mean	σ	Max	Mean	σ	Max
	fastPHASE	47.167	228.691	3514.098	213.997	715.714	14691.690
ER	Haplorec	49.082	256.980	4056.641	244.815	959.971	33365.050
	PTG	1.272	1.036	6.360	3.383	2.404	13.050
	fastPHASE	132.625	221.294	3253.907	140.402	236.700	3447.451
SE/SD	Haplorec	202.898	309.159	3884.019	895.172	723.861	4254.132
	PTG	152.855	246.303	3527.708	161.277	265.404	3814.880

4   Final Discussion and Conclusions

In this work we applied linear regression to predict errors in different Haplotype Inference algorithms. From the 7 biological properties considered, CSV was the most relevant, being used in five of the six models tested by ANOVA. CVS tries to capture conservation information on the sequences, which is an important indicator for the biological Parsimony principle. Of the three Haplotype Inference algorithms considered, only PTG applies a Pure Parsimony approach. For the other two methods, parsimony is not central.

The other variables selected for the regression models in the Individuals Base, $NS0$, $NS1$, $NB2$, and DHZ are indirectly related to variable HTZ, which was discarded by method Step-Wise from all models in this base.

The experiments show statistical evidence that all the regression models in Individuals can efficiently infer errors in HI Algorithms, except for response variable ER-fastPHASE in the SNPs Base. On the other hand, when the relative error is analyzed, a relatively high error and deviation can be observed. For the SNPs Base, the only model to present a reasonable result was the one for response variable ER-PTG.

An analysis of the shape of the values distribution showed that in Individuals, the shape approaches a normal, while in SNPs that is not necessarily the case. That could compromise the reliability of the statistical tests in the SNPs base. We compared the accuracy of these predictions with that of NN models previously developed, and the results demonstrate that regression is far more accurate than NN for the error prediction in HI algorithms. The quest for better SNPs Base predictors remains a challenge.

Due to time constraints, not all possible combinations of properties were analyzed as input for the neural network models. In the future, we intend to apply statistical analysis in order to identify the subset of the properties that are more likely the most relevant to estimate the Regression Models and to train the Neural Networks. Since there are many potential methods available [14], an extensive study will be required before selecting the ones that would yield the most useful predictive models.

Acknowledgments. This work was developed with financial support from Brazilian sponsoring agencies CNPq (processes 308572/2009-2 and 480927/2010-3), and FACEPE (process no. PBPG-0070-1.03/10), which the authors gratefully acknowledge.

References

1. Brown, D., Harrower, I.: Haplotyping as perfect phylogeny: Conceptual framework and efficient solutions. IEEE/ACM Trans. Comput. Biol. Bioinform. 3, 141–154 (2006)
2. Clark, A.: Inference of haplotypes from pcr amplified samples of diploid populations. Journal of Molecular Biology and Evolution 7, 111–122 (1990)
3. Consortium, T.I.H.: The international hapmap consortium. Nature 426, 789–796 (2003)
4. Ding, Z., Filkov, V., Gusfield, D.: A Linear-Time Algorithm for the Perfect Phylogeny Haplotyping (PPH) Problem. In: Miyano, S., Mesirov, J., Kasif, S., Istrail, S., Pevzner, P.A., Waterman, M. (eds.) RECOMB 2005. LNCS (LNBI), vol. 3500, pp. 585–600. Springer, Heidelberg (2005)
5. Eronen, L., Geerts, F., Toivonen, H.: Haplorec: efficient and accurate large-scale reconstruction of haplotypes. BMC Bioinformatics 7, 542 (2006)
6. Gusfield, D.: Inference of haplotypes from samples of diploids populations: Complexity and algorithms. Journal of Computational Biology 8, 305–323 (2001)
7. Gusfield, D.: Haplotyping as perfect phylogeny: Conceptual framework and efficient solutions. In: International Conference on Research in Computational Molecular Biology (RECOMB), pp. 166–175 (2002)
8. Gusfield, D.: Haplotype Inference by Pure Parsimony. In: Baeza-Yates, R., Chávez, E., Crochemore, M. (eds.) CPM 2003. LNCS, vol. 2676, pp. 144–155. Springer, Heidelberg (2003)
9. Halldórsson, B.V., Bafna, V., Edwards, N., Lippert, R., Yooseph, S., Istrail, S.: A Survey of Computational Methods for Determining Haplotypes. In: Istrail, S., Waterman, M.S., Clark, A. (eds.) SNPs and Haplotype Inference. LNCS (LNBI), vol. 2983, pp. 26–47. Springer, Heidelberg (2004)
10. Lancia, G., Pinotti, C.M., Rizzi, R.: Haplotype haplotyping populations by pure parsimony: Complexity of exact and approximation algorithms. INFORMS J. Computing 16, 348–359 (2004)
11. Li, Z., Zhou, W., Zhang, X.S., Chen, L.: A parsimonious tree-grow method for haplotype inference. Bioinformatics 21, 3475–3481 (2005)
12. Lin, S., Cutler, D.J., Zwick, M.E., Chakravarti, A.: Haplotype inference in random population samples. Am. J. Hum. Genet. 71(5), 1129–1137 (2002)

13. Montgomery, D., Runger, G.: Applied statistics and probability for engineers, 4th edn. LTC (2003)
14. Murtaugh, P.A.: Performance of several variable-selection methods applied to real ecological data. Ecology Letters 12(10), 1061–1068 (2009)
15. Niu, T., Qin, Z.S., Xu, X., Liu, J.S.: Bayesian haplotype inference for multiple linked single-nucleotide polymorphisms. Am. J. Hum. Genet. 70, 157–169 (2002)
16. Rosa, R.S., Guimarães, K.S.: Insights on Haplotype Inference on Large Genotype Datasets. In: Ferreira, C.E., Miyano, S., Stadler, P.F. (eds.) BSB 2010. LNCS, vol. 6268, pp. 47–58. Springer, Heidelberg (2010)
17. Rosa, R.S., Santos, R.H.S., Guimarães, K.S.: Accurate prediction of error in haplotype inference methods through neural networks. In: Proc. of the IJCNN 2012 (2012)
18. Scheet, P., Stephens, M.: A fast and flexible statistical model for large-scale population genotype data: applications to inferring missing genotypes and haplotypic phase. Am. J. Hum. Genet. 78(4), 629–644 (2006)
19. Stephens, M., Smith, N., Donnelly, P.: A new statistical method for haplotype reconstruction from population data. Am. J. Hum. Genet. 68, 978–989 (2001)
20. Sun, S., Greenwood, C.M., Neal, R.M.: Haplotype inference using a bayesian hidden markov model. Genet. Epidemiol. 31, 937–948 (2007)
21. Team, R.D.C.: R: A Language and Environment for Statistical Computing. R Foundation for Statistical Computing, Vienna, Austria (2011)
22. Wu, L., Zang, J., Chan, R.: Improved approach for haplotype inference based on markov chain. Lecture Notes in Operations Research, pp. 204–215 (2008)

Design and Implementation of ProteinWorldDB

Sérgio Lifschitz[1], Carlos Juliano M. Viana[1], Cristian Tristão[1],
Marcos Catanho[2], Wim M. Degrave[2], Antonio Basílio de Miranda[3],
Márcia Bezerra[3], and Thomas D. Otto[4]

[1] PUC-Rio - Departamento de Informática - Rio de Janeiro (RJ)
`sergio@inf.puc-rio.br`
[2] Laboratório de Genômica Funcional e Bioinformática
Instituto Oswaldo Cruz - Rio de Janeiro (RJ)
[3] Laboratório de Biologia Computacional e Sistemas
Instituto Oswaldo Cruz - Rio de Janeiro (RJ)
[4] Wellcome Trust Sanger Institute - Hinxton - United Kingdom

Abstract. This work involves the comparison of protein information in
a genomic scale. The main goal is to improve the quality and interpretation of biological data, besides our understanding of biological systems
and their interactions. Stringent comparisons were obtained after the
application of the Smith-Waterman algorithm in a pair wise manner to
all predicted proteins encoded in both completely sequenced and unfinished genomes available in the public database *RefSeq*. Comparisons
were run through a computational grid and the complete result reaches
a volume of over 900 GB. Consequently, the database system design is
a critical step in order to store and manage the information from comparisons' results. This paper describes database conceptual design issues
for the creation of a database that represents a data set of protein sequence cross-comparisons. We show that our conceptual schema and its
relational mapping enables users to extract relevant information, from
simple to complex queries integrating distinct data sources.

Keywords: Database design, Pairwise alignment, Complete genomes.

1 Introduction

The availability of complete genome sequences of numerous organisms, combined
with the computational progress occurred in the last few decades, provides an
opportunity to use holistic approaches in the detailed study of the genome structure, as well as gene prediction and functional classification.

Among these approaches, we are mainly interested in the comparative genome
analysis (or comparative genomics). It consists in the analysis and comparison
of genetic material from diverse species (or strains), aiming at investigating their
internal organization and evolution of the compared genomes (and the corresponding species). In addition, we are looking forward to revealing the function
of genes and non-coding regions in these genomes.

M.C.P. de Souto and M.G. Kann (Eds.): BSB 2012, LNBI 7409, pp. 144–155, 2012.
© Springer-Verlag Berlin Heidelberg 2012

This work reports results of the PWD[1] research project. It is an initiative dedicated to the comparison of protein information on a genomic scale. The goal is to improve the quality and interpretation of biological data, consequently, our understanding of biological systems and their interactions. Stringent comparisons were obtained after the application of the Smith-Waterman (SW) algorithm[15] in a pair wise manner to all predicted proteins encoded in both completely sequenced and unfinished genomes available at *RefSeq*[2] database.

Rigorous dynamic programming algorithms, such as SW, ensure the determination of the optimal alignment between pairs of sequences. In our case, we have run an implementation of the SW algorithm. However, due to their computational complexity, these algorithms are usually not suitable for the comparison of a large set of sequences. Therefore, we have considered distributed computing resources provided by the World Community Grid[3][12] to determine the sequence similarity level among almost 4 million proteins.

The result data available has reached a huge volume - over 900GB - of data, requiring a database system support. This is a fundamental factor if one wants to maximize the knowledge generation from the results yielded by the PWD project. Indeed, among others, we must consider data persistency, high availability and efficient access, all typical database technology features. Consequently, the database conceptual design becomes a relevant step to achieve the intended goals. Moreover, the corresponding logical (e.g. relational) schema would avoid performance bottlenecks, enable efficient querying and database maintenance.

This paper discusses conceptual modeling issues regarding the PWD project and all related database systems. We propose an extended conceptual schema in order to enable simple and complex queries, involving data obtained in our experiments and other relevant data sources, such as genomic sequences and taxonomies. We give then an overview of the database implementation issues, from the creation of the relational schema into a PostgreSQL DBMS up to samples of queries that define a formal access to our database.

It should be noted that there are many different research initiatives focusing on data modeling for bioinformatics (e.g. [6–8, 10, 11, 13, 16]). However, either there is no actual related project and the solutions proposed are not applicable to specific situations, or the conceptual models are so particular that we could not directly consider here. Furthermore, most approaches in the literature prioritize a systems view rather than a conceptual view. We claim that the modeling choices presented enforce the biological concepts and data integration issues.

This paper is organized as follows: we discuss next the motivation and the actual context of our research work, specifically some processing requirements. In Section 3 we discuss important issues with respect to conceptual data modeling, including some pros and cons of our modeling choices. Section 4 describes our relational schema and presents an overview of implementation and query answering. Finally, Section 5 concludes with future and ongoing work.

---

[1] www.proteinworlddb.org

[2] http://www.ncbi.nlm.nih.gov/RefSeq/ (version 21).

[3] http://www.worldcommunitygrid.org

2   Motivation

Comparative genomics comprehends the comparison of two or more genomes, including genomic sequences and their predicted protein content, the relative positions of their genes and other genomic context features that may be of functional (or regulatory) importance. It also includes the study of gene structure and organization, the presence and location of repetitive sequences, polymorphisms and several other characteristics that may help to differentiate genomes[9].

A detailed analysis of the predicted protein contents of an organism is an important step in genome analysis, and it has been applied to several studies with different objectives. Cancer, for instance, is a class of diseases where modifications in the expression pattern of several genes confer new biological properties to the cell. A better understanding of these alterations may provide new insights for the development of diagnostic and treatment procedures.

An important task is the identification of all protein-coding genes and their location in the genome sequence, as well as the characterization of their functions. Genomic sequences are scanned, searching for protein-coding genes, using computational gene models. For each new genome, each predicted gene is conceptually translated into a protein sequence; the predicted collection of protein sequences is the predicted proteome of the organism. Each predicted protein is used as a query sequence in similarity searches against repositories of biological sequences. Significant matches are added to the genomic sequence together with the gene position and its product description. More sophisticated methods for the search of gene families are also used for annotation. Collectively, these methods provide predictions for the proteome of a newly sequenced organism[9].

Additional information about a proteome can be obtained through the comparison of the set of protein sequences against itself, which identifies *paralogs* (genes originated after duplication events), through the comparison among different proteomes for the identification of *orthologs* (genes originated after speciation events), by studying fusion or fission events or new domain arrangements, and by studying the evolution of cellular, metabolic and regulatory functions.

Genomic analyses presents important computational challenges and one fundamental step is the efficient storage and management of the information derived from DNA and protein sequences, alignments, functions and locations of genes, protein families and domains, relations between genes of different organisms and chromosomal rearrangements, among others. The database system must be logically organized in such a way that all types of information are readily accessible and may be rapidly fetched, even for a large volume of data.

In what follows, we discuss the results of our practical comparisons, our first and straightforward conceptual model to represent it and the extended model that involves many other relevant information that enable a rather complete view of the biological experiment.

3   Database Conceptual Modeling

In our experiments, a set of 3,812,663 proteins from *RefSeq* version 21- consisting of all predicted proteins encoded in 458 completely sequenced and unfinished genomes - and 254,609 proteins from Swiss-Prot version 51.5[4] were compared, in a pair wise manner, with the program SSEARCH[14]. We have configured SSEARCH with standard parameters, and an E-value cut-off equal to one.

query gi, subject gi, SW score, bit score, e-value, % identity, alignment length, query start, query end, subject start, subject end, query gaps, subject gaps

67523787,67540134,2166,488.8,2.6e-138,0.336,1320,35,1275,67,1367,79,19

Fig. 1. A PWD match example report. The first line is the header of the listed values. The numbers are the ones that are stored.

For each significant match, a report is generated containing the identification of the pair of sequences compared and the alignment. The output format is given in Figure 1. A pair of protein sequences satisfying the required conditions to be stored was called a *hit*. A hit is defined by identifiers of the two sequences compared (Figure 1 has hit example between sequences *query_gi* = 67523787 and *subject_gi* = 67540134), and stores the validation measures of the pair wise comparison, besides additional information about the alignment, like similarity and coverage. The resulting matrix contains only hit information. The alignment itself was not stored.

Our main problem here was to define a database system that would help us for future querying and general data accesses. The goal is to store the results obtained in such a way that one could use these data together with other external data sources and generate relevant information. However, the whole system must consider the usual impedance mismatch among users offering a simple rather complete way to obtain the required information.

Figure 2 presents a first conceptual schema that can be output directly from the results. We have represented it with a conventional Entity- Relationship (ER) diagram, including min-max cardinalities. There are actually 3 possible combinations of hits involving translated ORFs and Proteins. All minimal cardinalities are zero as not all pair wise comparisons generate significant hits.

Results stored at the initial matrix contain only sequence identifiers and alignment information. The first step of this conceptual design was the creation of an entity that characterizes protein sequences. Information about the catalogued proteins compared in PWD includes the protein definition, its length and organism, and possible external references as protein identifiers (*RefSeq* and/or SwissProt). As the database must be kept up to date and updates occur, we

---

[4] http://www.uniprot.org/

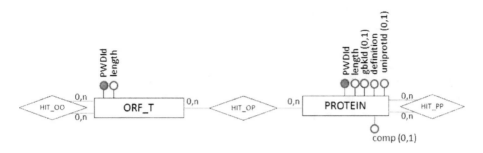

Fig. 2. First approach for a conceptual schema

identify those Proteins that have participated in registered comparisons. These are the main attributes of the Protein entity.

The amino acid sequences *translated_ORF* are represented by another entity *ORF_T* because they do not possess external identifiers. Information about these sequences includes the source organism, genomic location and length. Three types of distinct relationships between, proteins and ORFs are defined:

1. *hit_OO* is result of a comparison between *ORF_T* and *ORF_T*;
2. *hit_OP* is result of the comparison between *ORF_T* and proteins derived from SwissProt (proteins derived from *RefSeq* were not compared with *ORF_T*);
3. *hit_PP* is result of the comparison of *RefSeq* proteins with *RefSeq* and SwissProt proteins.

These relationships possess attributes that specify the pair wise results of PWD: $query_{gi}$, $subject_{gi}$, *SW score* (brute score of the comparison), *bit score* (normalized score), *e-value* (alignment significance), *%identity, alignment length*, *query_start, query_end, subject_start, subject_end, query_gaps, subject_gaps*.

However, we have some general and specific goals with this PWD project that cannot be solved only with the SSEARCH results and the corresponding output. There is a need of external data sources if one wants to check on the availability and feasibility. With respect to comparative genomics, hits represent only the result of protein-related genome comparisons. Further interesting questions depend on the protein coding gene physical position at its genomic context. For instance, the structure, organization and their genes relative position (gene order), and many other genomic features that may be of functional importance.

Therefore, Figure 3 gives an overview of our particular extended conceptual schema. We will discuss some of the data modeling alternatives and our design choices, which have guided us until the current conceptual schema.

We must observe that genes, transcripts, ORFs and genomic sequences are nucleotide sequences, while proteins and translated ORFs are amino acid sequences. The relationships between proteins and nucleotide sequences are constructed based on information from *RefSeq* version21.

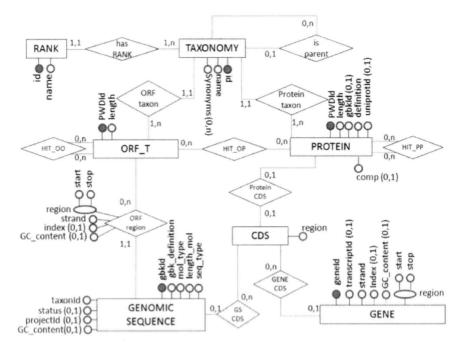

Fig. 3. A Conceptual Diagram for PWD Project

Conceptual Model Objects

We may explain entities, relationships and attributes in Figure 3 in more details. A protein is generated from a gene, which is a genomic sequence region. A protein coding gene is transcribed and produces a primary transcript that, after some processing, generates a mature transcript containing the protein coding sequence (CDS). The mature transcript is formed by the concatenation of sub-sequences containing information for the protein (exons) and untranslated regions (UTRs).

An ORF is a series of nucleotides codons extending up to the first termination codon. ORFs may not code for proteins. This way, all coding sequences (CDS) are ORFs but not all ORFs are coding for proteins.

Entity *Protein* represents the amino acid sequence of the protein that is related with the nucleotide sequence of CDS, and CDS with the gene and the genomic sequence containing it, keeping only an external reference to the transcript. Thus, *CDS* is an entity whose basic property is to keep up with the relationship between entities *Protein, Gene* and *Genomic Sequence*. This is done through the positioning of a given gene coding regions (exons) in the coordinate system of the genomic sequence that contains it. Each exon in a gene corresponds to a CDS sub-sequence, defined by an initial and a final position mapped into the coordinate system of the genomic sequence.

The entity *Gene* possesses an NCBI identifier - Entrez Gene[5], the region of its genomic sequence, defined by a start and a stop position, and a reading sense, its order in relation to the other genes in the genomic sequence, a transcript identifier (from *RefSeq*), and the GC-content. An amino acid sequence *ORF_T* refers to the genomic nucleotide sequence through an *ORF_region* delimited by a start and a stop position, inside the genomic sequence, with the reading sense, its position with respect to neighboring genes and the *GC-content*.

The *Genomic Sequence* entity possesses a *RefSeq* identifier, definition and sequence length, the type of organic molecule (DNA/RNA), status, type of sequence (chromosome, organelle, plasmid), an optional identifier of the respective genome project, GC content and an identifier of the source taxon.

Taxonomies and Classification

The classification used in PWD project is the same as the NCBI Taxonomy Database[2]. Each entry in the NCBI database is a *taxon*, also referred to as a *node*. The *root* node (taxid1) is at the top of the hierarchy. The path from the root node to any other particular *taxon* in this database is called its "lineage"; the collection of all of the nodes beneath any particular taxon is called its *subtree*.

In the conceptual model, the organism from which the genomic sequence was derived is the leaf node, defining the sequenced species (or an inferior rank like strain). It contains the taxID identifier from NCBI (a stable unique identifier for each taxon), the scientific and common names and synonyms. Each tree node has a rank, a parent node and may have descendent nodes. The taxonomic lineage may be obtained through a tree traversal from leaf nodes up to the root.

Conceptual Design Issues

We have first considered a database system exclusively oriented for the PWD project. Thus, the idea was to consider an entity called *Seq_AA* that would represent all compared amino acid sequences including annotated proteins and translated ORFs. This entity would relate with the hits matrix, and we would be able to specialize *Seq_AA* with either *RefSeq* or SwissProt as attributes. This entity would also be limited to sequences compared within the PWD project.

Within this particular conceptual schema, the amino acid sequences would relate with the nucleotide sequence through the entity *Coding_region*, and the latter with its source nucleotide sequence entity *Seq_NT* source. The problem of this representation is that it would be artificially adapting a biological concept, as ORFs were considered even if not coding regions. Moreover, another entity, *Seq_NT* source, also presented a wrong concept by dealing equally with both genomic and transcript sequences.

We have discussed some alternatives and decided to adopt the conceptual model depicted in Figure 3 due to the following reasons and modeling challenges:

- It is important to enable the database system to support updates as new genome sequences become available. It would be an error to limit the database only to the GCP project.

---

[5] http://www.ncbi.nlm.nih.gov/entrez/query.fcgi?db=gene

– The ORFs-T sequences, as a group of artificial (possible) proteins, brought many design problems. It becomes clear now that the sequences do not share the same characteristics with proteins, and need to be represented by and independent entity.
– Even if proteins with different origins could have distinct characteristics, with a conceptual viewpoint they all could be grouped as a single entity - Protein.

In our project, there are proteins that are originated from genomic projects including gene annotations, mRNA and CDS; those whose origins are only mRNA and proteins that are directly obtained from its sequencing process, without any reference to its original nucleotide. It brings another challenge for conceptual modeling and we have decided to model 3 types of amino acid sequences with specific characteristics and distinct research goals:

1. Protein sequences derived from finished genome projects (possessing the relationships genomic sequence gene transcript - CDS - protein) which will be the sequences considered in the comparative studies of proteomes, because they represent the complete predicted proteome of an organism;
2. Other protein sequences that may have these relationships are useful for the identification and validation of procedures and annotation results;
3. The experimental group of *ORF_T* will be used to evaluate the coding potential of these small ORFs, usually neglected by automatic gene prediction methods. These sequences were derived from complete prokaryotic genomic sequences. Therefore, the relationship *ORF_T* genomic sequence is mandatory, and validated sequences (with a high probability of being coding) may be included in the proteome comparative studies involving the database.

4   Relational Implementation and Queries

We have mapped our conceptual schema to a relational logical schema. Figure 4 shows the *Hits* similarity mapping. The recursive relationship *Hits* among *ORF_T* entity type participants were mapped for the table *hit_oo*. We also mapped recursive relationship *Hits* of *Proteins* (*hit_pp* table) and the relationship *Hits* among *Protein* and *ORF_T* (*hit_op* table).

Figure 5 shows the taxonomy relational mapping. The recursive relationship *is parent* is mapped to *taxonomy* table by the foreign key *taxonomy_id*. The relationships *have* among *ORF_T* and *PROTEIN* are mapped by the attribute *taxonomy_id* into *orf_t* and *protein* tables. We identify the attribute *synonymous* as a composite attribute. Then we have created the *synonymous* table that has different names for the same taxonomy.

Figure 6 shows the resulting logical schema diagram for our central dogma conceptual. Finally, we show in Figure 7 the complete logical schema diagram.

Queries: Validating the Relational Logical Data Model

With our relational schema in mind (Figure 7), it is possible to show how some relevant queries, involving most database objects, may be solved:

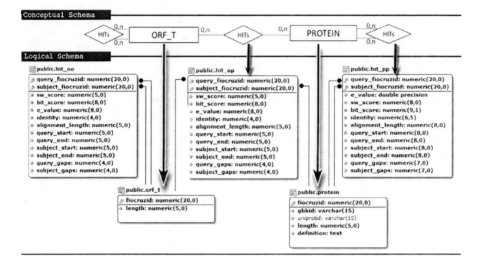

Fig. 4. Hits similarity mapping between translated ORFs and proteins

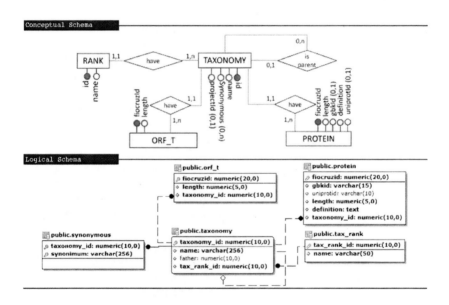

Fig. 5. Taxonomy mapping

1. Counting proteins in the database?

```
SELECT COUNT(DISTINCT p.fiocruzid)
FROM protein p
JOIN hit_pp_qid h ON p.fiocruzid = h.query_fiocruzid;
```

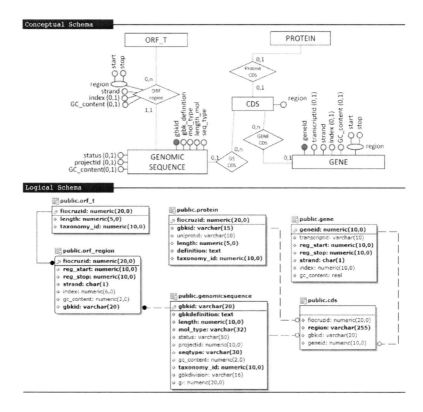

Fig. 6. Central Dogma Mapping

2. Proteins represented with genomic sequences?

```
SELECT p.fiocruzid, p.gbkid, p.definition
FROM genomic_sequence gs
JOIN cds ON gs.gbkid = cds.gbkid
JOIN gene g ON cds.geneid = g.geneid
JOIN protein p ON cds.fiocruzid = p.fiocruzid
JOIN hit_pp h ON h.query_fiocruzid = p.fiocruzid
```

3. How many genomes belong to a given (e.g. Vertebrata) taxonomy group?

```
SELECT gs.gbkid, gs.gbkdefinition, t.name
FROM genomic_sequence gs
JOIN taxonomy t ON gs.taxonomy_id = t.taxonomy_id
WHERE t.name LIKE '%Vertebrata%' ;
```

4. Return all hits above cut-off for a given protein X.

```
SELECT p.fiocruzid, p.definition, h.e_value, h.bit_score, h.sw_score
FROM hit_pp h
JOIN protein p ON h.subject_fiocruzid = p.fiocruzid
```

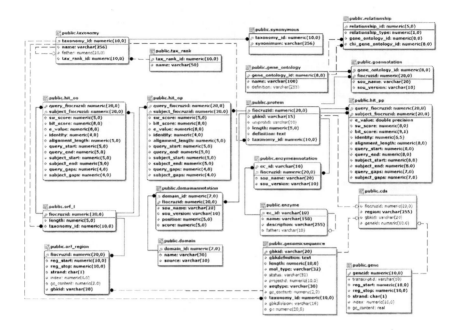

Fig. 7. A Logical Diagram for PWD Project

```
WHERE p.fiocruzid = 10957467
    AND (h.query_fiocruzid = 10957467 OR h.subject_fiocruzid = 10957467)
    AND h.e_value < 1.0e-5;
```

Complex queries are also relatively easy to follow if one considers our logical model depicted in Figure 7. Due to space limitations the reader is invited to check our project website.

5   Conclusions

We have discussed in this paper some conceptual modeling issues, including data modeling and queries, with respect to comparisons of protein information in a genomic scale. Due to the resulting data volume, the database system design becomes a critical step in order to extract significant information. We have discussed also implementation issues regarding our logical relational schema. We have implemented the logical model into PostgreSQL [3] as the underlying DBMS. Indexes access structures have been implemented to optimize some of the requested database queries.

Our main contributions rely on a general database system framework to represent sequence comparisons and the corresponding information with a fundamental and conceptual approach. This paper show an instantiation of our database schema considering data from the ProteinWorldDB project. We claim that many distinct queries, either simple or complex, may be stated in a straightforward manner. The system has been implemented and its first version is already available to the public.

Many different loading scripts were developed and will become available to the public. Our ongoing and future work involve annotation procedures and external data sources, such as Pfam [5] for protein domains, KEGG [1] (metabolic pathways) and controlled vocabulary based upon GeneOntology[4]. We are also tuning our database system in order to support complex queries and additional procedures, such as the identification of unique genes, *paralogs*, *orthologs* and many others.

Acknowledgements. We wish to thank IBM®, World Community Grid™ for their support.

References

1. KEGG: Kyoto Encyclopedia of Genes and Genomes,
   http://www.genome.jp/kegg/
2. NCBI Taxonomy Database, http://www.ncbi.nlm.nih.gov/Taxonomy/
3. PostgreSQL, http://postgresql.org
4. The Gene Ontology, http://www.geneontology.org/
5. The Pfam Protein Families Database, http://pfam.sanger.ac.uk
6. Chen, J.Y., Carlis, J.V.: Genomic data modeling. Information, Special issue: Data Management in Bioinformatics 28, 287–310 (2003)
7. Elmasri, R., Ji, F., Fu, J., Zhang, Y., Raja, Z.: Modelling Concepts and Database Implementation Techniques For Complex Biological Data. International Journal of Bioinformatics Research and Applications 3, 366–388 (2007)
8. Keet, C.M.: Biological Data and Conceptual Modelling Methods. Journal of Conceptual Modeling (2003)
9. Mount, D.: Bioinformatics: Sequence and Genome Analysis. Cold Spring Harbor Laboratory Press (2004)
10. Navathe, S.B., Kogelnik, A.M.: The Challenges of Modeling Biological Information for Genome Databases. In: Chen, P.P., Akoka, J., Kangassalu, H., Thalheim, B. (eds.) Conceptual Modeling. LNCS, vol. 1565, pp. 168–182. Springer, Heidelberg (1999)
11. Nelson, M.R., Reisinger, S.J., Henry, S.G.: Designing databases to store biological information. BIOSILICO 1, 134–142 (2003)
12. Otto, T.D., Catanho, M., Tristão, C., Bezerra, M., Fernandes, R.M., Elias, G.S., Scaglia, A.C., Bovermann, B., Berstis, V., Lifschitz, S., de Miranda, A.B., Degrave, W.: ProteinWorldDB: Querying radical pairwise alignments among protein sets from complete genomes. Bioinformatics (2010)
13. Pastor, O.: Conceptual Modeling Meets the Human Genome. In: Li, Q., Spaccapietra, S., Yu, E., Olivé, A. (eds.) ER 2008. LNCS, vol. 5231, pp. 1–11. Springer, Heidelberg (2008)
14. Pearson, W.: SSearch. Genomics 11, 635–650 (1991)
15. Smith, T., Waterman, M.: Comparison of Biosequences. Advances in Applied Mathematics 2, 482–489 (1981)
16. Zhou, X., Song, I.Y.: Conceptual Modeling of Genetic Studies and Pharmacogenetics. In: Gervasi, O., Gavrilova, M.L., Kumar, V., Laganá, A., Lee, H.P., Mun, Y., Taniar, D., Tan, C.J.K. (eds.) ICCSA 2005, Part III. LNCS, vol. 3482, pp. 402–415. Springer, Heidelberg (2005)

A Comparative Analysis of Public Ligand Databases Based on Molecular Descriptors

Ana T. Winck[1], Christian V. Quevedo[1], Karina S. Machado[2],
Osmar Norberto de Souza[1], and Duncan D. Ruiz[1]

[1] Pontificia Universidade Catolica do Rio Grande do Sul, GPIN-LABIO, PPGCC,
Porto Alegre, Brazil
{ana.winck,osmar.norberto,duncan.ruiz}@pucrs.br,
christian.quevedo@acad.pucrs.br
[2] Universidade Federal do Rio Grande, C3, Centro de Ciencias Computacionais,
Rio Grande, Brazil
karina.machado@furg.br

Abstract. A wide range of public ligand databases provides currently dozens of millions ligands to users. Consequently, exaustive *in silico* virtual screening testing with such a high volume of data is particularly expensive. Because of this, there is a demand for the development of new solutions that can reduce the number of testing ligands on their target receptors. Nevertheless, there is no method to reduce effectively that high number in a manageable amount, thus becoming this issue a major challenge of rational drug design. This article presents a comparative analysis among the main public ligand databases by measuring the quality and variations in the values of the molecular descriptors available in each one. It aims to help the development of new methods based on criteria that reduce the set of promising ligands to be tested.

Keywords: public ligand databases, molecular descriptors, virtual screening.

1 Introduction

The pharmaceutical industry is constantly trying to increase the rate with which it delivers new drugs to the market [1]. Although many technologies contribute to improve this process, it still takes about 10 or 15 years [2], and its associated costs are about 1,2 billion dollars [3]. Due to these reasons, there are current efforts to reduce time and costs, and to increase the quality of the candidate drugs. In the last decade, the structure-based virtual screening has achieved promising results, and the increasing availability of biological structures resulted in an important growth of the application of this technology. Ligand databases are essentially repositories able to store information related to small molecules. An ideal ligands database is the one that contains the largest possible number of information, provides easy access to such information, supplies fast queries response, and provides its data in a format reachable for the greatest number

M.C.P. de Souto and M.G. Kann (Eds.): BSB 2012, LNBI 7409, pp. 156–167, 2012.
© Springer-Verlag Berlin Heidelberg 2012

of computational systems [4]. Currently, we can find a significant number of these databases available for scientific community. Moreover, this number has substantially increased in the last years, as can be observed by the Nucleic Acids Research (NAR) reports [5].

The different ligand databases have their own particularities. In face of this, a careful analysis must be done to understand which molecular properties are stored and how these information are generated for each ligand. Another issue regards the type of access to these databases, considering whether they are public or private. Private ligand databases hold licenses with a high cost, which can make its acquisition unfeasible for many research groups [4]. The Cambridge Structural Database (CSD), a private one, currently stores about 500 thousand structures. All structures available in this database were analyzed using X-ray or neutron diffraction techniques [6]. However, its high cost licensing have motivated many research groups to use public databases.

In face of the enormous volume of data stored in these ligand databases, several researchers have applied virtual screening methods to select a subset of testing ligands in molecular docking experiments. Among such methods, we can notice a growing use of QSAR (Quantitative structure-active relationship) [7] models. These models are based on the identification of mathematical relationships between biological activities and chemical structures provided by molecular descriptors. This article presents a comparative analysis among the main public ligand databases by measuring the quality and variations in the values of the molecular descriptors available in each one. It aims to help the development of new methods based on criteria that reduce the set of promising ligands to be tested *in silico*.

2   Methods

2.1   Ligand Databases

The evaluation of all public ligand databases is complex and time-consuming. A comprehensive literature review was done to identify which ones are the most reputable for large research groups. The following public databases were found as the best ones.

- ChemBank [8]: it stores information on numerous small molecules and several of biomedically relevant assays.
- ChemDB [9, 10]: it was built using digital catalogs of many vendors and other public sources containing currently more than 5 million compounds stored. ChemDB supports multiple molecular formats updated periodically and includes a user-friendly graphical interface with different search capabilities.
- MMsINC Database [11]: it contains over 4 million non-redundant chemical compounds in 3D formats.
- NCI Database [12]: it is a web-based graphical user interface developed to allow quick searches by different criteria in more than 250,000 structures.

- PubChem [13]: it is a database to provide and encourage access within the scientific community to the most up to date and comprehensive source of chemical structures of small organic molecules and information on their biological activities. This database stores both compound information from the literature as well as screening probe data from Molecular Libraries Program.
- ZINC [4]: it is a database of commercially available compounds for virtual screening. It contains over 20 million purchasable compounds ready to docking. The same as ChemDB, it has a graphical interface to search for ligands with different parameters.

2.2   Target Protein and Ligands

Our target receptor is the InhA enzyme from *Mycobacterium tuberculosis* (MTB) [17]. It is a target to MTB because its inhibition disrupts the biosynthesis of mycolic acids, which is important to the arrangement of the mycobacterium cell wall [18]. In this work, the ligand databases were compared considering three ligands potentially inhibitors for InhA enzyme: Isoniazid (INH) [14], Triclosan (TCL) [15] and Ethionamid (ETH) [16].

INH has been the base of tuberculosis treatment for more than 50 years, since it was found as an effective agent [14]. Due to the resistance to INH in TB treatment, different inhibitors have been used, like Triclosan and Ethionamid. TCL is considered a good inhibitor candidate for this enzyme [15] as well as ETH, that only inhibits InhA activity when covalently linked to NADH, forming the adduct NADH-ETH [19].

The target receptor was studied in our previous work, where we have investigated data mining approaches to analyze data from fully flexible-receptor molecular docking experiments, looking for receptor snapshots to which a particular ligand binds more favorable [20–25].

3   Evaluation of Public Ligand Databases

Ligand databases provide molecular coordinates and a set of molecular properties. These properties are obtained by methods capable of capturing encoded information from the molecular structure and convert them into one or more useful numerical values. These values, which characterize the molecules properties, are also called molecular descriptors [26, 27].

Based on this information, a comparative study was done aiming to identify if the molecular properties made available in a database for a given ligand have the same values in the other databases. Thus, we selected a set of molecular descriptors that exist in common among the databases defined in this article. In the last few decades, this set of molecular descriptors has been used as one of the methods to perform an initial virtual screening in these large libraries of ligands. [28]. The set of descriptors selected to perform the comparison refers to methods based on molecular connectivity, and the following molecular descriptors collected:

- predicted octanol-water partition coefficient (*LogP*);
- number of hydrogen bond donors (*HBD*);
- number of hydrogen bond acceptors (*HBA*);
- molecular weight (*Mwt*);
- number of rotatable bonds (*NRB*).

As described in Methods Section, we considered three ligands as potentially inhibitors for InhA to compare the ligand databases: Isoniazid (INH), Triclosan (TCL) and Ethionamid (ETH). Information about this set of ligands were obtained from search tools in each database website, using as search string the SMILES code obtained from DrugBank [29]. Tables 1, 2 and 3 show results from searches made for INH, TCL and ETH, respectively.

Table 1. Feature description for Isoniazid molecule

Database	LogP	HBD	HBA	Mwt	NRB
ChemBank	-0.81/-0.51	2	3	137.139	1
ChemDB	-0.63/-0.82	3	3	137.139	2
MMsINC	-0.31	2	3	137.142	
NCI	-0.81	2	3	137.140	1
PubChem	-0.70	2	3	137.139	1
ZINC	-0.97	3	4	137.142	1

Table 2. Feature description for Triclosan molecule

Database	LogP	HBD	HBA	Mwt	NRB
ChemBank	Not available				
ChemDB	4.75/4.96	1	2	289.541	2
MMsINC	5.14	1	1	289.545	
NCI	Not available				
PubChem	5.00	1	2	289.541	2
ZINC	5.13	1	2	288.545	2

Table 3. Feature description for Ethionamide molecule

Database	LogP	HBD	HBA	Mwt	NRB
ChemBank	1.50/1.53	1	2	166.243	2
ChemDB	0.92/0.79	2	1	166.244	2
MMsINC	1.27	1	2	166.248	
NCI	1.52	1	1	166.240	2
PubChem	1.10	1	2	166.243	2
ZINC	1.46	2	2	166.249	2

As shown in such Tables, ChemBank and ChemDB databases provided more than one information for the partition coefficient (LogP). This has occurred because these databases use more than one method to predict LogP. Besides these differences, the databases presented different information for each ligand. The most representative variation regards the Partition Coefficient, number of hydrogen bond donors and acceptors and number of rotatable bonds.

3.1 Evaluation of Data Source

An investigation about methods used to populate each database was made to understand the origin of this variation. Results indicate that although several databases make use of OpenEye [30] software to predict each ligand 3D coordinates, they also make use of distinct methods to calculate molecular descriptors. This variety can be one factor implying in the distinct values found, since a ligand 3D structure can be identical, but having different atomic coordinates and 3D spaces. To evaluate this hypothesis, the information from each database must be submitted to the same software to calculate molecular descriptors.

Currently, there are several software able to calculate molecular descriptors of a ligand. In this study, we used the Dragon 5.5 software, which is capable of calculating up to 3,224 descriptors [31, 32]. This software receives molecular files in formats mol, mol2 and sdf, which are formats available by most databases of small molecules, and generates the values for the molecular descriptors that have been defined. However, to submit molecules to this software and to evaluate this procedure is a time-consuming process. Because of this, experiments were made for the same three ligands and, for each ligand being tested, files provided for databases were captured in their distinct format (mol2, mol and sdf), accordingly to each website availability. The files were submitted to Dragon [31, 32] to calculate the same properties detailed in the previous Tables 1, 2 and 3.

Tables 4, 5 and 6 present results obtained from Dragon software. The file of the TCL ligand obtained from Chembank was downloaded, although available information was in a 2D format, and this format is not suitable to be submitted to Dragon. Besides, this ligand was not found in BD NCI.

By analyzing Tables 4, 5 and 6 it is possible to notice that the values of molecular descriptors generated by software Dragon have converged. Thus, comparing the information provided by databases (Tables 1, 2 and 3) with the information generated by software Dragon (Tables 4, 5 and 6) we conclude that the variations

Table 4. Feature description for Isoniazid molecule obtained from Dragon software

Databases	LogP	HBD	HBA	Mwt	NRB
ChemBank	-0.81	3	4	137.160	1
ChemDB	-0.81	3	4	137.160	1
MMsINC	-0.81	3	4	137.160	1
NCI	-0.81	3	4	137.160	1
PubChem	-0.81	3	4	137.160	1
ZINC	-0.81	3	4	137.160	1

Table 5. Feature description for Triclosan molecule obtained from Dragon software

Databases	LogP	HBD	HBA	Mwt	NRB
ChemBank	Not available				
ChemDB	5.11	1	2	289.540	2
MMsINC	5.11	1	2	289.540	2
NCI	Not available				
PubChem	5.11	1	2	289.540	2
ZINC	5.11	1	2	289.540	2

Table 6. Feature description for Ethionamide molecule obtained from Dragon software

Databases	LogP	HBD	HBA	Mwt	NRB
ChemBank	1.53	1	2	166.270	2
ChemDB	1.53	2	2	166.270	2
MMsINC	1.53	2	2	166.270	2
NCI	1.53	2	2	166.270	2
PubChem	1.53	2	2	166.270	2
ZINC	1.53	2	2	166.270	2

found among these tables occur since different software were used for predicting molecular descriptors.

This evaluation presents an important consideration, since nowadays many researchers have used pre-filtering techniques based on characteristics of ligands that have an optimal affinity with the receptor of interest. That is, software is used to calculate the molecular descriptors of these ligands with optimal affinity and subsequently applying these criteria in the databases. The resulting set of ligands is used to perform experiments that are more sophisticated.

Thus, the variations presented in this paper are a warning about the limitations in using some pre-filtering techniques. Depending on the combination of prediction software of molecular descriptors and the database to be used, the variations can be significant. For example, if the values of molecular descriptors obtained by software Dragon 5.5 (Tables 4, 5 and 6) were selected and researched as filtering criterion in their respective databases, only Ethionamide, located in ChemBank, would be found.

4   Results and Discussion

Besides the verification of data sources and how similar are the information available in these databases, a study to evaluate the particularities offered from each database was done by considering a set of features as follows:

- **Allows downloading data sets.** Download the information of ligands for a local basis allows the execution of experiments on a large scale;
- **Type of search.** Provides the possibility of finding a specific molecule or select a set of ligands with some characteristic in common;

- **Test of structural similarity among ligands.** Applications of structural similarity based techniques are widely used in the scientific community;
- **Supply molecules prepared for molecular docking software execution.** It is faster when the atomic charge of the molecule is already calculated because it is necessary to carry on molecular docking tests, in order to validate these tests;
- **Total volume of ligands.** Databases that stores a greater number of ligands, even having files with variations 3D about the same ligand, have more chances to obtain promising results;
- **Format files.** The format provided is important due to the limitations of reading files that still exist in many biological software tools.

After establishing these features, we performed a bibliography search and verification about current information available in each database website. Table 7 shows a summary of such evaluation.

Table 7. Summary of main features of the ligand databases being considered

	ChemBank	ChemDB	MMsINC	NCI	PubChem	ZINC
Download Subsets	x	x	x	x	x	x
Download all data	x	x		x	x	x
Search by name	x	x	x	x	x	x
Search by SMILE	x	x	x	x	x	x
Search by structure	x		x	x		x
Search by substructure	x	x	x	x	x	x
Serch by molecular descriptors	x	x	x	x	x	x
Search by similarity	x	x		x	x	x
Ready for docking						x
Total volume of ligands (in million)	4.5	5	4	0.25	32	20
File formats available	sdf	mol, mol2 sdf, PDB	mol, sdf PDB	mol2, sdf	ASN1, XML sdf	mol2, sdf flexibase

Through the analysis of Table 7 it is possible to notice that PubChem is the database presenting the greater number of ligands available. In addition, it is the only one informing the volume of the ligand. However, there is no information about the method used to calculate the volume. The main weakness found in this database is the lack of preparation for molecular docking software execution, which implies in time consuming. It would be necessary to previously set the charge of all atoms for each ligand available and, then, to perform molecular docking experiments.

ZINC is the second bigger database regarding the amount of ligands and the only one among the surveyed databases that provides in its files the information about partial load, which is accepted by the main software for molecular docking.

The ChemBank, ChemDB and MMsINC databases have some features that are not present in PubChem, for instance the search by exact structure. On the other hand, the amount of compounds is relatively low, ranging from 4 to 5 million ligands. Besides, MMsINC does not allow the download of all available ligands and both ChemDB and PubChem do not allow the search by the exact structure.

The NCI database possesses a little more than 250 mil structures. It's possible to note this database hosts a very limited number of ligands, according to Table 7. Bigger databases have over than 32 million compounds. However, data stored in the NCI refer to existing drugs, making these ligands as reliable results.

4.1   Lessons Learned

Strengths and weaknesses of the characteristics of each database were identified. However, only ZINC has met all criteria shown in Table 7 and, in the next section, we describe in more detail this database.

Characterization of ZINC Database. ZINC is a well-known database being constantly updated, originally published in 2005. Currently, this database is in version 12 and provides more than 20 million molecules. Due to its high amount of data, several subsets were developed to provide an easier access on more specific data. Each subset has many rules to be applied as feature filters. The main subsets are listed below.

- Lead-Like: based on Teague et al. [33] rules to identify lead compounds;
- Fragment-like: this subset rules are described in Carr et al [34];
- Drug-like: based on rule of 5 from Lipinski [35], but has many compounds that are exception;
- All-purchasable: subset containing all molecules that can be obtained from suppliers;
- Everything: subset with all molecules, including those ones that cannot be obtained;

In the case of users do not find a subset meeting their needs, ZINC allows users to build their own subset to test experiments, and such subsets become available for one week. Other facility is that information about the number of similar molecules are available for each subset, classifying groups in 50%, 60%, 70%, 80%, 90%, 95% and 99% of accuracy according to Tanimoto similarity [36].

This investigation has been motivated aiming at identifying the source of data stored in ZINC. The ligands built to populate this database were generated by a set of software from OpenEye [30]. These software collect information about 2D structures from small molecules suppliers, in a sdf format and, making use of prediction, provide information about 3D structures as results. Finally, CO-RINA software was used to calculate the molecular physicochemical properties, which are molecular weight, polar solvation energy, polar surface area, number of hydrogen bond donors and acceptors, number of rotatable bonds, partition coefficient and atomic charge [36].

After identifying physicochemical properties, a selection is done discarding those ligands with values greater those ones pre-determined by the database. Table 8 presents the main restriction of this database, where ligands that have some restriction presented in such table are discarded. It's worth to mention that there are some exception, as including drugs that violates these restrictions [4].

Table 8. Properties used by ZINC to discard ligands

Properties	Restrictions
Molecular weight	> 700 g/mol
LogP	> 6 or < - 4
Number of hydrogen bond donors	> 6
Number of hydrogen bond acceptors	> 11
Number of rotatable bonds	> 15
Atom type	All atom types except: H, C, N, O, F, S, P, Cl, Br, I

The website of this database contains search tool to query stored data. If the searched molecule is found it can be downloaded in a mol2, sdf and Flexibase formats. If the search is for a set of molecules, there are a series of filter of physicochemical features to specify the required set. To look for similar molecules the ZINC or SMILES code need to be inserted or even draw the required molecule in the molecular editor. In summary, this database stores information about atomic coordinates, partial atomic charges of each chemical element, the nine physicochemical properties, data about molecule suppliers and a set of molecule identification attributes. These attributes are both international pattern codes to designate molecules (SMILES and SMARTS) and ZINC codes to identify the molecules.

5   Conclusions

Currently there are several repositories of ligands available. However, there is a lack of comparative studies among these databases in order to assist researchers in identifying which is the best data repository to be used in her process of virtual screening. In this paper, a comparison between six major databases was presented in detail. This comparison was separated in two steps.

In the first step, the study consisted of submitting the same ligand to the six ligand databases. This process was reapplied with 3 pro-drugs known for the particular case of InhA, an enzyme in which our research group has a deeper knowledge. This study identified a number of variations between the values of molecular descriptors available in each database. To investigate the origin of these variations, experiments were performed with the software Dragon 5.5. Results of this evaluation showed that such variations occur due to the databases use different software tools to predict their molecular descriptors. The main consequence on these variations refers to retrieve different sets of ligands when pre-filtering the database using, as parameters, the molecular descriptors.

These variations affects directly approaches that use criteria based on molecular descriptors. For example, Drug-like properties, also known as the Lipinski rules [28], evaluate the oral bioavailability of a compound. This rule is based on 4 molecular descriptors (LogP, HBD, HBA, and MWT) and using these criteria, a ligand can be either selected or discarded. Thus, a given database can infer that the ligand X meets the criteria of "rule of five" and in another database the same ligand cannot fit in the Lipinski rules. Thus, conforming the way molecular descriptors values are obtained becomes necessary, avoiding the divergences between databases.

In the second step, a study was performed to evaluate the peculiarities on each database, highlighting strengths and weaknesses on the characteristics of each database. At the end of the evaluation, ZINC was the only database to cover all criteria set out in this work. Later, we presented important details of this database that is the second largest with regards to the number of available ligands. A great advantage of this repository is that its provided files already have the partial atomic charges assigned, ready to be executed in molecular docking software. Thus, when comparing the major public database of ligands, ZINC satisfied the main properties established in this study for the virtual screening process, being a very promising database in the *in silico* research of new drugs.

Despite noting the existing variations in values of molecular descriptors among the databases, it is known that a study with a larger number of samples is important even to perform more refined statistical inferences. Thus, a challenge to be solved is the development of scripts to perform the search of the ligands automatically, by seeking for canonical SMILES code in each database. There are some restrictions to be solved, for example, existing variations of pH.

However, with these structures identified by the same SMILES code, it would be possible to submit these ligands to be evaluated by the same software to calculate their corresponding descriptors, making an inference in a wider range of ligands, and generating the statistical analyzes of variations obtained. A more specific evaluation would be the check of sets of ligands that would meet Lead-Like [33], Fragment-Like [34], and Drug-Like [35] in different databases.

Acknowledgements. This work was supported by grants from MCT/CNPq to DDR and to ONS. ONS is CNPq Research Fellow. CVQ is supported by FAPERGS/CAPES PhD scholarship.

References

1. Lyne, P.D.: Structure-based virtual screening: an overview. Drug Discovery Today 7, 1047–1055 (2002)
2. Caskey, C.T.: The drug development crisis: efficiency and safety. Annual Review of Medicine 58, 1–16 (2007)
3. Kapetanovic, I.: Computer-aided drug discovery and development (CADDD): *in silico*-chemico-biological approach. Chemical Biology Interaction 171, 1047–1055 (2008)

4. Irwin, J.J., Shoichet, B.K.: ZINC – a free database of commercially available compounds for virtual screening. Journal of Chemical Information and Modeling 45(1), 177–182 (2005)

5. Cochrane, G.R., Galperin, M.Y.: The 2010 nucleic acids research database issue and online database collection: a community of data resources. Nucleic Acids Research 38, D1–D4 (2009)

6. Allen, F.H.: The Cambridge Structural Database: a quarter of a million crystal structures and rising. Acta Crystallographica Serie B 58(3), 380–388 (2002)

7. Hansch, C., Leo, A.: Exploring QSAR: fundamentals and applications in chemistry and biology. ACS, Washington (1995)

8. Seiler, K.P., George, G.A., Happ, M.P., Bodycombe, N.E., Carrinski, H.A., Norton, S., Brudz, S., Sullivan, J.P., Muhlich, J., Serrano, M., Ferraiolo, P., Tolliday, N.J., Schreiber, S.L., Clemons, P.A.: ChemBank: a small-molecule screening and cheminformatics resource database. Nucleic Acids Research 36, D351–D359 (2008)

9. Chen, J.H., Linstead, E., Swamidass, S.J., Wang, D., Baldi, P.: ChemDB update: full-text search and virtual chemical space. Bioinformatics 23(17), 2348–2351 (2007)

10. Chen, J.H., Swamidass, S.J., Dou, Y., Bruand, J., Baldi, P.: ChemDB: a public database of small molecules and related chemoinformatics resources. Bioinformatics 21(22), 4133–4139 (2005)

11. Masciocchi, J., Frau, G., Fanton, M., Sturlese, M., Floris, M., Pireddu, L., Piergiorgio Palla, P., Cedrati, F., Rodriguez-Tomé, P., Moro, S.: MMsINC: a large-scale chemoinformatics database. Nucleic Acids Research 37, D284–D290 (2009)

12. Ihlenfeldt, W.-D., Voigt, J.H., Bienfait, B., Oellien, F., Nicklaus, M.C.: Enhanced CACTVS Browser of the Open NCI Database. Journal of Chemical Information and Modeling 42, 46–57 (2002)

13. Austin, C.P., Brady, L.S., Insel, T.R., Collins, F.S.: Molecular biology: NIH Molecular Libraries Initiative. Science 306(5699), 1138–1139 (2004)

14. Middlebrook, G.: Sterilization of the tubercle bacili by isonicotinic acid hidrazide and the incidence of variants resistant to the drug in vitro. American Review of Tuberculosis 65, 765–767 (1952)

15. Kuo, M.R., Morbidoni, H.R., Alland, D., Sneddon, S.F., Gourlie, B.B., Staveski, M.M., Leonard, M., Gregory, J.S., Janjigian, A.D., Yee, C., Musser, J.M., Kreiswirth, B., Iwamoto, H., Perozzo, R., Jacobs, W.R., Sacchettini, J.C., Fidock, D.A.: Targeting Tuberculosis and Malaria through Inhibition of Enoyl Reductase: Compound Activity and Structural Data. Journal of Biological Chemistry 278, 20851–20859 (2003)

16. Baulard, A.R., Betts, J.C., Engohang-Ndong, J., Quan, S., McAdam, R.A., Brennan, P.J., Locht, C., Besra, G.S.: Activation of the pro-drug ethionamide is regulated in mycobacteria. Journal of Biolological Chemistry 275, 28326–28331 (2000)

17. Dessen, A., Quémard, A., Blanchard, J.S., Jacobs Jr., W.R., Sacchettini, J.C.: Crystal structure and function of the isoniazid target of Mycobacterium tuberculosis. Science 267, 1638–1641 (1995)

18. Schroeder, E.K., Norberto de Souza, O., Santos, D.S., Blanchard, J.S., Basso, L.A.: Drugs that inhibit mycolic acids biosynthesis in *Mycobacterium tuberculosis*. Current Pharmaceutical Biotechnology 3, 197–225 (2002)

19. Banerjee, A., Dubnau, E., Quemard, A., Balasubramanian, V., Um, K.S., Wilson, T., Collins, D., de Lisle, G., Jacobs Jr., W.R.: InhA, a gene encoding a target for isoniazid and ethionamide in *Mycobacterium tuberculosis*. Science 263, 227–230 (1994)

20. Machado, K.S., Schroeder, E.K., Ruiz, D.D., Wink, A., Norberto de Souza, O.: Extracting Information from Flexible Receptor-Flexible Ligand Docking Experiments. In: Bazzan, A.L.C., Craven, M., Martins, N.F. (eds.) BSB 2008. LNCS (LNBI), vol. 5167, pp. 104–114. Springer, Heidelberg (2008)
21. Winck, A.T., Machado, K.S., Norberto-de-Souza, O., Ruiz, D.D.D.: FReDD: Supporting Mining Strategies through a Flexible-Receptor Docking Database. In: Guimarães, K.S., Panchenko, A., Przytycka, T.M. (eds.) BSB 2009. LNCS (LNBI), vol. 5676, pp. 143–146. Springer, Heidelberg (2009)
22. Machado, K.S., Winck, A.T., Ruiz, D.D., Norberto de Souza, O.: Discretization of Flexible-Receptor Docking Data. In: Ferreira, C.E., Miyano, S., Stadler, P.F. (eds.) BSB 2010. LNCS (LNBI), vol. 6268, pp. 75–79. Springer, Heidelberg (2010)
23. Machado, K.S., Winck, A.T., Ruiz, D.D., Norberto de Souza, O.: Mining flexible-receptor docking experiments to select promising protein receptor snapshots. BMC Genomics 11, 1–13 (2010)
24. Machado, K.S., Winck, A.T., Ruiz, D.D., Norberto de Souza, O.: Mining flexible-receptor docking data. WIREs DMKD 1, 532–541 (2011)
25. Machado, K.S., Schroeder, E.K., Ruiz, D.D., Cohen, E.M.L., Norberto de Souza, O.: FReDoWS: a method to automate molecular docking simulations with explicit receptor flexibility and snapshots selection. BMC Genomics 12, S6 (2011)
26. Portugal, J.: Evaluation of molecular descriptors for antitumor drugs with respect to noncovalent binding to DNA and antiproliferative activity. BMC Pharmacol. 9, 11 (2009)
27. Todeschini, R., Consonni, V.: Handbook of Molecular Descriptors, 668 p. Wiley-VCH, Weinheim (2000)
28. Charifson, P.S., Walters, W.P.: Filtering databases and chemical libraries. Journal of Computer-Aided Molecular Design 16, 311–323 (2002)
29. Wishart, D.S., et al.: DrugBank: a comprehensive resource for *in silico* drug discovery and exploration. Nucleic Acids Research 34, D668–D672 (2006)
30. OpenEye. OpenEye Scientific Software (February 2012),
    http://www.eyesopen.com/
31. Mauri, A., Consonni, V., Pavan, M., Todeschini, R.: DRAGON software: an easy approach to molecular descriptor calculations. MATCH 56, 237–248 (2006)
32. Talete srl. Dragon for Windows version 5.5 (January 2011),
    http://www.talete.mi.it/
33. Teague, S.J., Davis, A.M., Leeson, P.D., Oprea, T.: The Design of Leadlike Combinatorial Libraries. Angewandte Chemie International Edition 38(24), 3743–3748 (1999)
34. Carr, R.A.E., Congreve, M., Murray, C.W., Rees, D.C.: Fragment-based lead discovery: leads by design. Drug Discovery Today 10(14), 987–992 (2005)
35. Lipinski, C.A.: Drug-like properties and the causes of poor solubility and poor permeability. Journal Pharmacological Toxicological Methods 44(1), 235–249 (2000)
36. Irwin, J.J.: Using ZINC to acquire a virtual screening library. Current Protocols on Bioinformatics 22(14), 14.6.1–14.6.23 (2008)

Phylogenetic Distance Computation Using CUDA

Wellington S. Martins[1], Thiago F. Rangel[2], Divino C.S. Lucas[3],
Elias B. Ferreira[1], and Edson N. Caceres[4]

[1] Institute of Computing, Federal University of Goias, Goiania, Brazil
{wellington,eliasferreira}@inf.ufg.br
[2] Department of Ecology, Federal University of Goias, Goiania, Brazil
thiagorangel@icb.ufg.br
[3] Institute of Computing, State University of Campinas, Campinas, Brazil
cesar@lsc.ic.unicamp.br
[4] Institute of Computing, Federal Univ. Mato Grosso do Sul, Campo Grande, Brazil
edson@dct.ufms.br

Abstract. Some phylogenetic comparative analyses rely on simulation procedures that use a large number of phylogenetic trees to estimate evolutionary correlations. Because of the computational burden of processing hundreds of thousands of trees, unless this procedure is efficiently implemented, the analyses are of limited applicability. In this paper, we present a highly parallel and efficient implementation for calculating phylogenetic distances. By using the power of GPU computing and a massive number of threads we are able to achieve performance gains up to 243x when compared to a sequential implementation of the same procedures. New data structures and algorithms are also presented so as to efficiently process irregular pointer-based data structures such as trees. In particular, a GPU-based parallel implementation of the lowest common ancestor (LCA) problem is presented. Moreover, the implementation makes intensive use of bitmaps to efficiently encode paths to the tree nodes, and optimize memory transactions by working with data structures that favors coalesced memory accesses. Our results open up the possibility of dealing with large datasets in evolutionary and ecological analyses.

Keywords: parallel computing, GPU, phylogenetics.

1 Introduction

The information contained in the topology of a phylogenetic tree is of maximum interest for biologists, as inferences of evolutionary processes are derived from the phylogenetic relationship between species. This relationship can be captured by the pairwise phylogenetic distance between species, and when multiple species are included in the analysis, such information can be conveniently stored in a squared distance matrix [1], the so called patristic distance matrix. A patristic distance is the sum of the lengths of the branches that link two species in a tree. Some of the statistical methods used to study evolutionary processes require

M.C.P. de Souto and M.G. Kann (Eds.): BSB 2012, LNBI 7409, pp. 168–178, 2012.

the raw or standardized patristic distance matrix (e.g. estimate phylogenetic diversity), whereas other methods require the transformation of the distance matrix into a variance-covariance matrix (e.g. linear regression analysis).

A variety of computational algorithms have been applied to phylogenetic analysis. However, the majority of these algorithms (e.g. Distance-matrix based, Maximum Parsimony, Maximum Likelihood, and Bayesian analysis) attempt to infer the phylogenetic relationships of the species under study, i.e. they construct a phylogenetic tree from a set of genes, species, or other taxanomic groups (taxa). These methods have been parallelized to be able to deal with larger and more accurate phylogenetic trees and to reduce the computational burden. Some of the proposals make use of computational grids, clusters and even GPUs [2–6]. Conversely, phylogenetic comparative analysis does not attempt to infer phylogenies but rather it uses phylogenetic trees to compare species. Although computational algorithms have been developed for comparative analyses, only recently simulations involving large data sets (typically 1,000 or more) have been used. These simulations are usually carried out using traditional phylogenetic comparative tools, that do not exploit novel parallel computer architectures [7–9].

The work presented in this paper aims at contributing to conduct such large scale phylogenetic comparative analyses. We present a parallel computing algorithm to calculate patristic distance matrices, commonly used in large scale statistical analyses. We show that by using a massive number of threads and efficient data structures we are able to achieve performance gains up to 243x when compared to the sequential implementation. In addition, intensive use of bitmaps and GPU specific optimizations contribute to efficiently process irregular pointer-based data structures such as trees. The rest of the paper is organized as follows. In Section 2 we present the algorithm and the CUDA implementation. The experimental results are shown and discussed in Section 3. Finally in Section 4 we present the conclusions and future works.

2 GPU Implementation

Before doing the actual parallel processing, the implementation has to take care of a few preprocessing tasks. The input data, a phylogenetic tree, is stored in a Newick formatted file, and it has to be read and parsed. The Newick format [10] is a simple way to represent phylogenetic trees in text form using parentheses and commas. Figure 1 shows a simple phylogenetic tree represented in this format and the corresponding tree. Each species at the tips of the tree has a name followed by a number indicating the length of the branch connecting it to its immediate ancestor. Species, or taxa, sharing an immediate common ancestor are wrapped by parentheses. In the example, the left innermost parentheses indicate that B and A are sister taxa. The next set of parentheses outward indicate that D is the sister to the C group. And so on.

We wrote a parser in C++ that reads the phylogenetic tree from the file and converts it to an internal representation. The parser does a reverse post-order traversal and process trees as they appear in the Newick format. That is, for the

(((A:2,B:3)C:4,D:2)E:3,(F:3,(G:2,H:1)I:5)J:4)K;

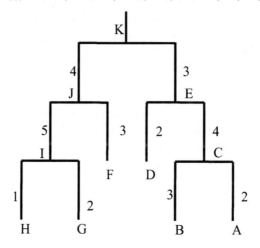

Fig. 1. *Newick format and the corresponding tree*

tree illustrated in Figure 1, the order will be A, B, C, D, E, F, G, H, I, J and K. This means that a child always comes before its parent, and in particular, that the root comes last. Furthermore, since the input phylogenetic tree may contain polytomies (nodes with more than two branches), the preprocessing has to treat these phylogenetic uncertainty by making sure that the resulting tree is a binary tree structure.

In order to favor coalesced memory accesses we store the information in a structure of arrays (SoA), where each array holds part of the information of the tree. Besides the node's name and its branch length, we also store the parent and the children of each node. In addition, we make sure that the information related to species and ancestors (internal nodes) are easily accessible by storing species information on the low index part of the arrays, and ancestors information on the high index part of the arrays. This way, a simple comparison of the array index with the number of species determines whether the index points to a species or an ancestor. We also use a negative index as a null pointer to easily find out whether an index refers to a species/ancestor or to null.

This representation is shown in Figure 2 for the tree presented previously in Newick Format. As can be seen from the figure, the species A, B, D, G, H and F were mapped to the six first positions of the array that represents the tree. Conversely, the ancestors K, E, C, J and I were mapped to the last five positions. Since the tree is binary, and all internal nodes always have two children, the number of ancestors will be one less of the number of species. Note that the value -2 is used to indicate that the node has no children whereas the value -1 indicates that the root (K) has no parent.

The patristic distance matrix is an n by n matrix that describes the phylogenetic similarity between all pairs of n species. This matrix is symmetric, with

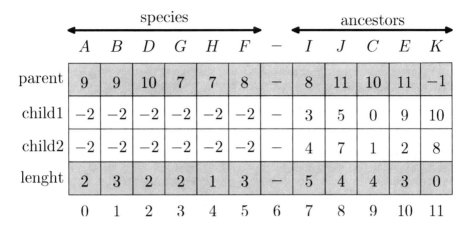

	species						ancestors				
	A	B	D	G	H	F	$-$	I	J	C	E	K
parent	9	9	10	7	7	8	$-$	8	11	10	11	-1
child1	-2	-2	-2	-2	-2	-2	$-$	3	5	0	9	10
child2	-2	-2	-2	-2	-2	-2	$-$	4	7	1	2	8
lenght	2	3	2	2	1	3	$-$	5	4	4	3	0
	0	1	2	3	4	5	6	7	8	9	10	11

Fig. 2. *Data representation example*

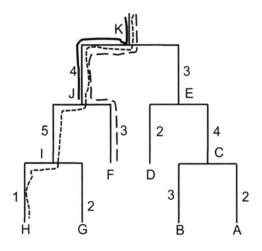

Fig. 3. *Distance calculation*

zeros on the diagonal, so we only need to calculate half (e.g upper right half) of the elements of the matrix. However, to calculate a single distance between a pair of species we need to sum up all branch lengths connecting the species. One way of calculating that is by first finding the lowest common ancestor (LCA), i.e, the lowest node in the tree that has both species as descendants. Then the distance between the species is given by the sum of the distances of each species to the root minus twice the distance of the LCA to the root. Figure 3 illustrates this for the species H and F. The LCA is J and its distance to the root is 4. The distances of H and F to the root are 10 and 7 respectively. So the distance between H and F is 10 plus 7 minus 8 (2 times 4), that is, 9.

To calculate each distance in the phylogenetic distance matrix we need to find the lowest common ancestor (LCA) between pairs of species, so this operation has to be efficiently implemented. There have been some algorithms proposed in the literature to calculate the LCA, including some parallel algorithms [11–13]. However, we do not know of any parallel implementation of LCA for GPUs. We propose a method that works in two steps. First, we calculate the distance of every node (species and ancestors) to the root of the tree. This is done by moving up the tree, following the parent link, and accumulating the branch lengths. In this process, we keep track of the direction we came from, i.e. right or left, in a path number that encodes the path taken in a binary form. This path number is used as a key, together with the accumulated distance value, to a hash table. By comparing two path numbers we are able to easily find out the path number of the LCA and, consequently, its distance to the root.

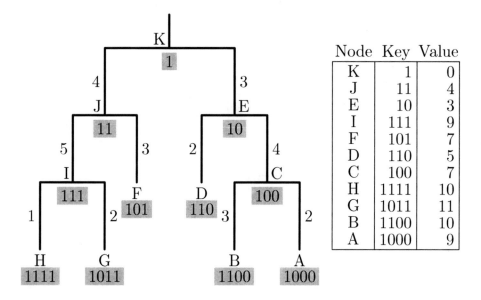

Fig. 4. *Path numbers (keys) and distances (values)*

To illustrate this process, consider again the species H and F in Figure 4. The path numbers for H and F are 1111 and 101 respectively. We assume that 0 indicates that we came from the right branch and 1 otherwise. Furthermore, an extra 1 is used as a prefix of the path number to indicate the root. Thus, the path number for the species F is set initially with 1 (prefix) and, as we move up, the bits are shifted left and additional bits are included, 0 for right and 1 for left, resulting in 101. To find out the lowest common ancestor of H and F, we compare both path numbers, 1111 and 101, starting from the least significant bit and stopping when the bits differ. The path number of the LCA will be equal to 1 (prefix) followed by the coincident bits, in this case 1. This results in the path number 11, which is the key for the distance value of node J. Figure 4 shows the path numbers (keys) and distances (values) for all nodes of the tree. The table shown in the figure is actually implemented as a hash table so that queries can be processed quickly. We opted for an open addressing implementation since we know the number of entries to be stored, i.e., two times the number of species minus 1. Collissions were treated by using quadratic probing and making sure the table size is prime and the load factor is less than 0.5. The whole process necessary to generate the patristic distance matrix was implemented on the following kernel.

```
1 Kernel
2 for all threads do in parallel
3   associate a thread with a species or ancestor
4   while parent link is not root do
5     accumulate distance and pathnumber
6     follow parent link
7   end of while
8   store distance and pathnumber in the hash table
9   sync_threads (barrier synchronization)
10  associate a thread with an element of the array of distances
11  for k = 1 to (number of species / 2) do
12    map array element to matrix
13    query hash table to find distances of the species to the root
14    compare the pathnumbers of the pair of species
15    find LCA and query hash table to find its distance to the root
16    calculate the distance between the pair of species
17      distance 1st + distance 2nd - 2 distance of LCA
18    store the distance in the array representing the matrix
19  end of for
20 end of for all
```

We now explain, step by step, how the previous method is implemented in parallel in the GPU. Since the distances between all pairs of species need to be calculated for each tree, we assign each phylogenetic tree to a block of threads and let the threads inside a block cooperate on the calculations of a distance matrix. This allows us to explore an additional level of parallelism, with many threads working on a single tree. To calculate the table shown in Figure 4 we associate each species, and ancestor, to a thread. As the threads follow the parent links they accumulate the distance to the root and build the corresponding path number [lines 3-7]. This calculation was implemented in two phases. First the threads process the species, in the first half of the arrays. Then, after a barrier synchronization, the ancestors (internal nodes) are processed. All nodes (species and ancestors) could be processed simultaneously, but that would require more threads per block. Depending on the size of the tree, we can adjust the group of working threads by doing a round robin scheduling.

After storing the path numbers and the corresponding distances to the root in the hash table [lines 8-9], the phylogenetic distance matrix can finally be calculated. As already mentioned, the matrix is symmetric, and we only need to calculate half of the elements of the matrix. However, the rows have different

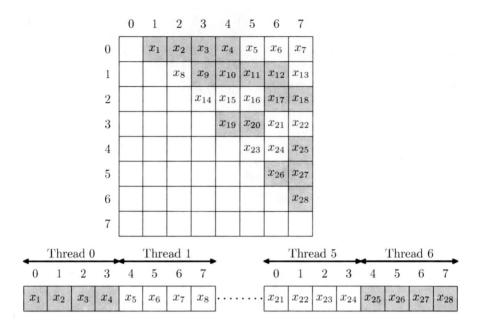

Fig. 5. *Distance matrix example*

numbers of elements and assigning each row to a thread would be inefficient. In order to balance the load, the upper right half of the matrix is physically stored in an array. Thus, each thread takes care of the same number of elements, in this case, the number of species divided by two [lines 10-11]. A function is used to map the indexes of the array to the corresponding indexes (row and column) of the matrix [line 12]. Figure 5 illustrates this scheme for a tree containing eight species. With the indexes (row and column) of the matrix, the thread can now query the hash table to find out the distances of the species up to the root, and the corresponding path numbers (keys). The path numbers are compared to determine the LCA of the two species. The index of the LCA is used to query the hash table to retrieve its distance to the root. Finally, the distance between the two species is calculated and the result stored in the array representing the matrix [lines 13-19].

3   Experimental Results

The experiments were conducted using an Intel Core2 Duo 1.6GHz, 4GB RAM, an NVIDIA Tesla C1060, and Linux (Ubuntu 11.04) operating system. The Tesla consists of an array of 30 Streaming Multiprocessors (SMs), each with eight scalar processor cores (total of 240 cores) with 1.3GHz of clock speed, 4 GB of global memory and 16 KB of shared memory. Our algorithms were programmed

in C/C++ and CUDA C/C++ 4.0. We used the following phylogenies to calculate the patristic distance matrices: Phyllostomidae (bats) [14] with 126 species,
Carnivora (mammals) [15] with 209 species, Hummingbirds [16] with 304 species,
and Amphibia (amphibians) [17] with 419 species.

Table 1. 100 Trees

Species	Phylocom	Serial	Parallel	
			Total	Transfer
126	0.36 (0.02)	0.33 (0.01)	0.02 (0.00)	0.01 (0.00)
209	0.98 (0.02)	0.87 (0.02)	0.02 (0.00)	0.02 (0.00)
304	4.07 (0.05)	1.99 (0.02)	0.03 (0.00)	0.02 (0.00)
419	8.81 (0.03)	4.01 (0.04)	0.04 (0.00)	0.02 (0.00)

Table 2. 1000 Trees

Species	Phylocom	Serial	Parallel	
			Total	Transfer
126	3.28 (0.04)	3.32 (0.02)	0.16 (0.01)	0.14 (0.01)
209	9.84 (0.15)	8.67 (0.04)	0.20 (0.01)	0.15 (0.01)
304	40.53 (0.36)	20.34 (0.30)	0.28 (0.01)	0.17 (0.01)
419	88.24 (0.45)	39.96 (0.41)	0.38 (0.01)	0.17 (0.01)

Table 3. 10000 Trees

Species	Phylocom	Serial	Parallel	
			Total	Transfer
126	32.69 (0.26)	32.96 (0.20)	1.55 (0.02)	1.35 (0.02)
209	95.98 (0.76)	86.83 (0.37)	1.90 (0.01)	1.41 (0.01)
304	406.13 (1.53)	200.35 (0.66)	2.58 (0.02)	1.46 (0.02)
419	879.32 (1.75)	397.27 (0.80)	3.61 (0.02)	1.60 (0.02)

We compared our results to the results produced by Phylocom, a well known
open source software for the analysis of phylogenetic community structure and
trait evolution. We also developed a serial version of our parallel algorithm for
the sake of comparison. The Phylocom calculates various metrics, including the
patristic distance matrix (option -phydist). The source code was modified to call
the distance matrix funcion as many times as necessary for the experiments;
input and output were counted only once. The same strategy was used for the
serial version developed. For the parallel implementation version, we replicated
the input tree and transfered the resulting structure to the GPU's global memory.
At the end of the processing, one of the resulting distance matrices is chosen
to be moved back to the CPU to check the calculations. The execution times

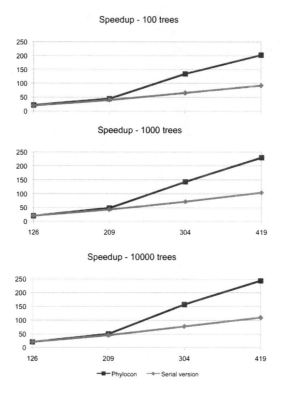

Fig. 6. *Speedups in relation to Phylocom and the Serial version*

(in seconds) obtained are shown in the tables 1, 2 and 3. The reported numbers are average of 10 independent runs; the standard deviations are shown between parentheses. In the last column of the tables is shown the time taken to transfer the data from the CPU to the GPU and vice versa. This time is approximately 87%, 74%, 58% and 44% of the total parallel runtime, for the number of species considered (126, 209, 304, 419), showing that as we increase the size of the trees, less time is spent on moving data to and from the GPU. However, even for small trees, the speedup is never less than 20, showing that is worth doing the calculation on the GPU.

The charts of Figure 6 show the speedup versus the number of species (126-Phyllostomidae, 209-Carnivora, 304-Hummingbirds and 419-Amphibia) for the Phylocom program and the serial version of the parallel implementation. For a given number of trees, for example 10.000 trees, as we increase the the number of species, the parallel version is 21, 50, 157, and 243 times faster than the serial Phylocom program. The distance matrix computation alone, without taking into account the transfer time, gives us a maximum speedup of 440x. So even dealing with an irregular pointer-based tree structure, the GPU is able to have an up to 243x performance advantage over a single CPU for large enough problem sizes. Similar behavior happens when the number of trees is 100 and 1.000. The benefit is lower with small number of trees primarily because the amount of parallelism is lower. For 10.000 trees, and 419 species (Amphibia) our kernel launchs 4.190.000 (10.000 * 419) threads. It takes the GPU just over 2 seconds (plus 1.6 seconds to transfer data) to calculate 10.000 distance matrices.

4   Conclusions and Future Works

A highly parallel and efficient GPU based implementation to support large scale simulations used in comparative phylogenetics has been proposed and evaluated in this paper. The solution proposed allows for the calculation of patristic distance matrices, a commonly used operation in large scale statistical analyses. It includes a GPU-based parallel implementation of the lowest common ancestor (LCA) problem and a load balanced strategy to better distribute the work among the participating threads.The use of new data structures, intensive use of bitmaps and GPU specific optmizations resulted in an efficient way to process irregular pointer-based data structures such as trees. We were able to achieve performance gains up to 243x when compared to a sequential implementation.

This work will be further improved by including a parallel strategy to random generate the phylogenetic trees used in large scale simulations. In addition, we plan on implementing statistical methods that require the patristic distance matrices as a starting point. Some of these methods produce a set of coefficients from the distance matrices. This is advantageous to the current implementation since only a small set of data needs to be returned to the CPU. In particular, we will investigate the use of Moran's I measure of spatial autocorrelation to be use in ecological analysis.

Acknowledgments. Thanks to Thierson Couto and Humberto Longo for insightful discussions about the patristic distance matrix calculation. EBF thanks CAPES for his MSc scholarship (Proc. 946684). TFR is supported by CNPq, grants 564718/2010-6, 474774/2011-2, 310117/2011-9.

References

1. Felsenstein, J.: Confidence limits on phylogenies: an approach using the bootstrap. Evolution 39(4), 783–791 (1985)
2. Suchard, M.A., Rambaut, A.: Many-core algorithms for statistical phylogenetics. Bioinformatics 25, 1370–1376 (2009)
3. Ayres, D.L., Darling, A., Zwickl, D.J., Beerli, P., Holder, M.T., Lewis, P.O., Huelsenbeck, J.P., Ronquist, F., Swofford, D.L., Cummings, M.P., Rambaut, A., Suchard, M.A.: BEAGLE: an application programming interface and high-performance computing library for statistical phylogenetics. Syst. Biol. 61(1), 170–173 (2012)
4. Stamatakis, A.: Parallel and Distributed Computation of Large Phylogenetic Trees. In: Zomaya, A.Y. (ed.) Parallel Computing for Bioinformatics and Computational Biology: Models, Enabling Technologies, and Case Studies. John Wiley & Sons, Inc., Hoboken (2005)

5. Petzold, E., Merkle, D., Middendorf, M., von Haeseler, A., Schmidt, H.A.: Phylogenetic Parameter Estimation on COWs. In: Zomaya, A.Y. (ed.) Parallel Computing for Bioinformatics and Computational Biology: Models, Enabling Technologies, and Case Studies, John Wiley & Sons, Inc., Hoboken (2005)
6. Williams, T.L., Bader, D.A., Moret, B.M.E., Yan, M.: High-Performance Phylogeny Reconstruction Under Maximum Parsimony. In: Zomaya, A.Y. (ed.) Parallel Computing for Bioinformatics and Computational Biology: Models, Enabling Technologies, and Case Studies. John Wiley & Sons, Inc., Hoboken (2005)
7. Martins, E.P.: COMPARE, version 4.6b. Computer programs for the statistical analysis of comparative data. Department of Biology, Indiana University, Bloomington, IN
8. Webb, C.O., Ackerly, D.D., Kembel, S.W.: Phylocom: software for the analysis of phylogenetic community structure and trait evolution. Bioinformatics 24, 2098–2100 (2008)
9. Fourment, M., Gibbs, M.: PATRISTIC: a program for calculating patristic distances and graphically comparing the components of genetic change. BMC Evol. Biol. 6, 1 (2006)
10. Olsen, G.: "Newick's 8:45" Tree Format Standard (1990),
    http://evolution.genetics.washington.edu/phylip/newick_doc.html
11. Aho, A., Hopcroft, J., Ullman, J.: On finding lowest common ancestors in trees. In: Proc. 5th ACM Symp. Theory of Computing (STOC), pp. 253–265 (1973)
12. Bender, M.A., Farach-Colton, M.: The LCA Problem Revisited. In: Gonnet, G.H., Viola, A. (eds.) LATIN 2000. LNCS, vol. 1776, pp. 88–94. Springer, Heidelberg (2000)
13. Schieber, B., Vishkin, U.: On finding lowest common ancestors: simplification and parallelization. SIAM Journal on Computing 17(6), 1253–1262 (1988)
14. Dartzmann, T., von Helversen, O., Mayer, F.: Evolution of nectarivory in phyllostomid bats (Phyllostomidae Gray, 1825, Chiroptera: Mammalia). Evolutionary Biology 10, 165 (2010)
15. Bininda-Emonds, O.R.P., Cardillo, M., Jones, K.E., MacPhee, R.D.E., Beck, R.M.D., Grenyer, R., Price, S.A., Vos, R.A., Gittleman, J.L., Purvis, A.: The delayed rise of present-day mammals. Nature 446, 507–512 (2007)
16. McGuire, J.A., Witt, C.C., Altshuler, D.L., Remsen Jr., J.V.: Phylogenetic systematics and biogeography of hummingbirds: Bayesian and maximum likelihood analyses of partitioned data and selection of an appropriate partitioning strategy. Systematic Biology 56, 837–856
17. Pyron, R.A., Wiens, J.J.: A large-scale phylogeny of Amphibia including over 2800 species, and a revised classification of extant frogs, salamanders and caecilians. Molecular Phylogenetics and Evolution 61, 543–583 (2011)

Exploring Molecular Evolution Reconstruction Using a Parallel Cloud Based Scientific Workflow

Kary A.C.S. Ocaña[1], Daniel de Oliveira[1,2], Felipe Horta[1],
Jonas Dias[1], Eduardo Ogasawara[1,2], and Marta Mattoso[1]

[1] COPPE/Federal University of Rio de Janeiro, Rio de Janeiro, Brazil
[2] CEFET-RJ/Federal Center of Technological Education, Rio de Janeiro, Brazil
`{kary,danielc,jonasdias,ogasawara,marta}@cos.ufrj.br,`
`fhorta@poli.ufrj.br`

Abstract. Recent studies of evolution at molecular level address two important issues: reconstruction of the evolutionary relationships between species and investigation of the forces of the evolutionary process. Both issues experienced an explosive growth in the last two decades due to massive generation of genomic data, novel statistical methods and computational approaches to process and analyze this large volume of data. Most experiments in molecular evolution are based on computing intensive simulations preceded by other computation tools and post-processed by computing validators. All these tools can be modeled as scientific workflows to improve the experiment management while capturing provenance data. However, these evolutionary analyses experiments are very complex and may execute for weeks. These workflows need to be executed in parallel in High Performance Computing (HPC) environments such as clouds. Clouds are becoming adopted for bioinformatics experiments due to its characteristics, such as, elasticity and availability. Clouds are evolving into HPC environments. In this paper, we introduce SciEvol, a bioinformatics scientific workflow for molecular evolution reconstruction that aims at inferring evolutionary relationships (*i.e.* to detect positive Darwinian selection) on genomic data. SciEvol is designed and implemented to execute in parallel over the clouds using SciCumulus workflow engine. Our experiments show that SciEvol can help scientists by enabling the reconstruction of evolutionary relationships using the cloud environment. Results present performance improvements of up to 94.64% in the execution time when compared to the sequential execution, which drops from around 10 days to 12 hours.

Keywords: Molecular Evolution Reconstruction, Scientific Workflow, Cloud.

1 Introduction

Over the last two centuries, the study of evolutionary relationships among species became an important issue. Since Charles Darwin formulated the theory of evolution, evolutionary studies evolved in a fast pace. Nowadays, many evolutionary studies are based on computer simulations, which are related to several bioinformatics programs that implement molecular evolution reconstruction (MER) methods. These simulations are increasing in scale and complexity due to the necessity of processing

M.C.P. de Souto and M.G. Kann (Eds.): BSB 2012, LNBI 7409, pp. 179–191, 2012.
© Springer-Verlag Berlin Heidelberg 2012

thousands of several available genomes [1]. MER experiments focus on the inference of evolutionary relationships among individuals, populations, species and higher taxonomic entities using molecular data. Managing MER experiments is far from trivial, since they are computing-intensive and process large amounts of data due to their exploratory characteristic. MER experiments are based on a pipeline of scientific programs, which may be modeled as scientific workflows [2]. Scientific workflows provide a structured view of the experiment, improving its design.

Scientific Workflow Managements Systems (SWfMS) are able to define, execute and manage workflows. Several SWfMS register the execution of the workflow through provenance data [3]. Provenance represents the ancestry of an object. Provenance of an object, such as a data product, contains information about the process used to derive the object, in this case the data product. It provides important documentation that is essential to preserve the data, to determine their quality and authorship, and to reproduce as well as to interpret and validate the associated scientific results generated by large scale scientific experiments. Provenance data can be queried in several ways finding which parameters led to specific results and is also used for verifying if a specific data file was indeed generated or which parameters were consumed. Particularly in MER workflows, during the evolutionary relationship inference, thousands of intermediate data files are produced. Scientists have to analyze each one of these files and associate their content to the activity they are related to and which parameters were consumed to produce them. Provenance automatically provides associations between parameters and data files. Due to the existing large amount of files, this task is susceptible to errors when performed manually.

MER workflows are exploratory by nature. Each MER workflow execution explores several different parameter values e.g. evolutionary codon substitution models to verify if the sites in genes are under positive Darwinian selection (explained in Section 2). Although there are some programs [4, 5] that execute evolutionary reconstruction, the parameter exploration is performed and managed manually, which may become a burden and error-prone task. Depending on the amount of input data, the multiple sequence alignment (MSA) method [6], the phylogenetic algorithm and the complexity of the codon substitution models (method proposed by Goldman and Yang [7]), each MER workflow execution may demand weeks or even months to produce results when executed sequentially (using a single machine). For example, a regular execution of a MER workflow using 132 multi-fasta files (with an average of 25 biological sequences in each file) of genomic data needs approximately 232.85 hours (10 days) to finish when executing in a regular desktop.

Since acquiring and maintaining a high performance computer demands a lot of investments and a permanent specialized supporting team, many scientists choose to work on desktops or small clusters (commonly Beowulf ones). This scenario leads to avoid exploring different codon substitution models in MER workflows since the total execution time can be prohibitive. These explorations may be feasible in High Performance Computing (HPC) environments, where parallelism techniques are applied to improve the performance of the workflow execution. Clouds have demonstrated applicability to a wide-range of problems in several scientific domains [8]. Elastic scaling of resources (hardware and software capabilities) is a key characteristic of clouds and scientists benefit from it. Cloud computing [9] recently started to provide HPC capabilities [10], which are needed in bioinformatics experiments. Another attractive cloud feature is that scientists do not have to operate or maintain a parallel computer cluster or a grid. This paper presents SciEvol, a cloud-based parallel scientific workflow that defines exploratory MER experiments. SciEvol is able to infer the evolution and to analyze the forces and mechanisms of evolutionary process, which thus have to test all possible variations of codon models.

SciEvol explores input data, estimating parameters and testing hypotheses for studying an evolutionary process. In this way, SciEvol enables to: (i) estimate synonymous and non-synonymous substitution rates for detecting purifying, neutral or positive evolution in protein-coding nucleotide sequences; and (ii) identify which selected sites can be, for example, under positive selection. SciEvol uses the codon substitution models implemented in the codeml program available in PAML package [11]. SciEvol has two sub-workflows, the comparative genomics procedure modeled in SciHmm [12] and phylogenetic analysis modeled in SciPhy [13]]. SciEvol is designed to take advantages of parallel execution in cloud environments. It uses a specialized cloud parallel workflow engine, named SciCumulus [14].

SciCumulus manages the parallel execution of SciEvol by applying parameter sweep [14] mechanisms and creating a virtual cluster formed by several Virtual Machines (VMs). Each VM processes independent activities, consuming different input data in parallel. All provenance data related to SciEvol is captured and managed automatically by SciCumulus. SciCumulus has already been successfully experienced in other complex computing-intensive bioinformatics experiments (e.g. SciPhy [13] and SciHmm [12]). Experimental results reinforce the importance of SciEvol to help scientists in their evolutionary inferences and further analysis with the help of provenance data. Experiments also confirm the benefits of using cloud environments, which enables performance gains of up to 94.64% in the execution time when compared to the sequential execution.

The remainder of this paper is organized as follows. Section 2 brings important evolutionary biology background. Section 3 describes the specification of the SciEvol workflow and discusses its implementation using SciCumulus parallel cloud workflow engine. Section 4 shows experimental results. Section 5 discusses related work and finally Section 6 concludes this paper.

2   Background on Adaptive Molecular Evolution Reconstruction

This section presents background on MER methodologies. It discusses about statistical methods for detecting molecular evolution by comparing synonymous and non-synonymous rates in protein-coding DNA sequences. To achieve this goal we have to perform a likelihood ratio test (LRT) [15] to identify evolutionary lineages under Darwinian selection or to infer critical amino acids in a protein under diversifying selection. One of the main purposes in the MER analysis is to determine which selective pressure is being exerted in biological sequences to understand the evolutionary behavior. Most commonly used cases of adaptive MER have been identified through comparison of synonymous (*i.e.* silent or d_S) and non-synonymous (*i.e.* amino acid-changing or d_N) substitution rates in protein-coding DNA sequences, thus providing case studies of natural selection in action on the protein molecule. A comparison regarding the fixation of the rates in these two types of mutations provides a powerful tool to understand the effect of natural selection in the evolution of molecular sequences. One metric that has been continuously used in several researches [16, 17] is the non-synonymous and synonymous substitution rate (*i.e.* $\omega = \frac{d_N}{d_S}$). The values of d_N and d_S are defined as the numbers of non-synonymous and synonymous substitutions *per* site and their rate ω represents the selective pressure at the protein level. The selection is determined by ω, as follows: (i) $\omega > 1$ for positive selection; (ii) $\omega < 1$ for purifying or negative selection and (iii) $\omega = 1$ for a scheme of neutral substitutions. This way, $\omega > 1$ (also called as positive Darwinian selection) means that the non-synonymous mutations offer advantages for proteins and have a higher probability of fixation than synonymous mutations [7].

The estimation of d_N and d_S using likelihood [7] is based on an explicit model of codon substitution (*i.e.* substitution rates). Nielsen and Yang [18] propose a method where the sites of a protein can assume distinct values of d_N and d_S. It uses several models that differs each one from the others according to the distribution of ω among the codons. Thus, the detection of positive selection can be performed individually for each one of the sites of a given protein. There is a large collection of codon substitution models (*branch*, *site* and *branch-site* models) [19]. SciEvol is based on *site models* method proposed by Nielsen and Yang [18], implemented by codeml program (explained in Section 3). Substitution codon models (M0, M1, M2, M3, M7, and M8) are used to determine how ω varies among sites [18].

Model M0 (one-ratio) assumes a single value of ω for all sites. Model M1 (nearly neutral) assumes one class of conserved sites ($0 < \omega < 1$) and one class of neutrally evolving sites ($\omega = 1$), distributed respectively in different proportions p_0 and p_1, where $p_1 = 1 - p_0$. Model M2 (positive selection) adds an additional site class of $\omega > 1$ with p_2 presenting values greater than 1 for d_N / d_S (named ω_2), to accommodate sites evolving under positive selection. Model M3 (discrete) assumes a general discrete model, with p_K frequencies and ω_K ratios for K site classes estimated as free parameters; it provides a significant improvement over M0, indicating significant variation in ω among sites. Model M7 uses a beta-distribution [20] of ω for data that is limited to the interval (0,1). M8 adds an additional site class of $\omega > 1$ to M7 [18]. At this point these codon models are fit to the dataset with codeml, which estimates parameters and calculates the log-likelihood values (*l*) for each codon model.

However, LRT statistics need to be calculated independently to compare the relative fit of the different hierarchically nested models. Models are nested [15] in cases where the most complex model (the one with a larger number of free parameters) is derived from a simpler one (with fewer free parameters), *i.e.*, the complex model contains at least one more parameter over the simpler one. The detection of positive selection of sites occurs if the complex model for the nested models contains certain proportion of sites with $\omega > 1$, whereas the simple model does not contains $\omega > 1$. When the LRT suggests the presence of sites with $\omega > 1$, we can use the Naïve Empirical Bayes (NEB) method to calculate the conditional (posterior) probability distribution of ω for each site given the data at the site [19] (as presented in Table 1).

The LRT statistic is based on the p-value for $2\Delta l$ statistical test (difference between the l at the nested models). This p-value is obtained using the chi-square distribution (X^2) [20] that allows the ascertain of statistical significance between the nested models, assuming degrees of freedom (simply d.f. or number of statistical parameters). The evaluation enables us to determine whether the more complex model provided a significantly better fit than the more restricted model. Model descriptions and model comparison strategies for these analyses are outlined in the literature [21, 22].

3   SciEvol: A Workflow for Molecular Evolution Reconstruction

This section presents the conceptual specification of SciEvol (Fig. 1), a scientific workflow for MER analysis. SciEvol is composed by 12 activities: (i) a Python script to format the data input by removing stop codons; (ii) MSA construction, (iii) MSA format conversion to the PHYLIP format [23]; (iv) a Python script to format the PHYLIP file; (v) phylogenetic tree construction; (vi-xi) evolutionary analysis execution, which is composed by six phases, executing codon substitution models M0, M1, M2, M3, M7, and M8; and (xii) evolutionary data analysis. The first activity formats the multi-fasta input file by deleting the stop codons in sequences. The second activity of SciEvol constructs the MSA using MAFFT program (or other MSA program such as ProbCons, ClustalW, Muscle, or Kalign). It receives a pre-formatted multi-fasta file as input and produces the MSA in FASTA format as output. The next activity executes ReadSeq to convert the MSA in FASTA format to PHYLIP format (referenced as "phylip-file-one" to be used in the phylogenetic tree construction activity). The fourth activity formats the input PHYLIP format file ("phylip-file-one") and generates a second PHYLIP format file (referenced as "phylip-file-two" to be used in the evolutionary analysis activity), adjusting the number of sites to be multiple of three (a limitation of codeml) and adding the parameter "I" at the ending of the first line of the file, due to format definition. The fifth activity is responsible for constructing the phylogenetic tree using RAxML. It receives the "phylip-file-one" as input and produces a phylogenetic tree in Newick format [23] as output.

MER model exploration is done from the sixth activity until the eleventh. These activities execute six different phases of the MER exploration (related to each one of the six codon substitution models) using codeml. As inputs it receives the phylogenetic tree, the "phylip-file-two" and the respective *codeml.ctl* file. Input contains the parameter values to be explored, for each one of the codon substitution models. As output, it generates a set of files containing evolutionary information. The last activity automatically processes the output files obtained from the previous activities by: (i) applying LRT on nested models, obtaining statistical results reported as a summary and (ii) representing sites under evolutionary codon models in graphics. After that, scientists already have available information to manipulate and to analyze results of large-scale genomic data in order to infer evolutionary hypotheses.

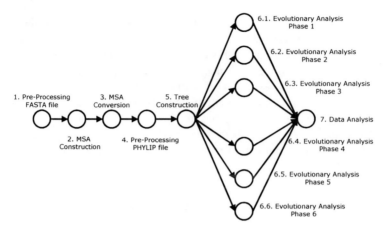

Fig. 1. SciEvol Conceptual View

During the course of a MER experiment there may be many analyses to be performed, which implies executing the SciEvol many times *i.e.* varying input data and parameters. These various executions of SciEvol may incur in a total execution time that cannot be processed in an acceptable time by a desktop or a small cluster. This way, we execute SciEvol in parallel in a HPC environment using SciCumulus. SciEvol was modeled in XML using SciCumulus as presented in Fig. 2. One of the advantages of using SciCumulus is that, for each parameter explored, SciCumulus automatically records all steps and files associated to the executed activities. These records can be queried using a high-level interface that allows for a systematic analysis of the experiment in a partial, or as a whole, after its completion. For additional information on SciCumulus please refer to Oliveira *et al.* [14].

```
<SciCumulus>
 <database name="scicumulus" server="mp4-5b.dyndns.info" port="5432"/>
 <SciCumulusWorkflow tag="SciEvol" description="MER" exectag="scievol" expdir="/scievol/">
  <SciCumulusActivity tag="MSA" templatedir="/scievol/template/" activation="./experiment.cmd">
   <Relation reltype="Input" name="rel_in_1" filename="input_step_1.txt"/>
   <Relation reltype="Output" name="rel_out_1" filename="output_step_1.txt"/>
   <File filename="experiment.cmd" instrumented="true"/>
  </SciCumulusActivity>
 </SciCumulusWorkflow>
</SciCumulus>
```

Fig. 2. An excerpt of SciEvol XML specification file in SciCumulus

4   Experimental Results

In this section we present a biological evaluation of the results achieved by executing SciEvol and highlight its benefits for testing MER, thus detecting positive Darwinian selection in protein-coding nucleotide sequences. Besides the biological results, we have also evaluated the performance and scalability of the parallel execution of SciEvol using SciCumulus engine. We have deployed all bioinformatics applications:

MAFFT, ReadSeq, RAxML, PAML, and SciCumulus components on the top of Amazon EC2 environment.

A. Cloud Environment Setup

Amazon EC2 provides different types of VM. Each VM has different memory and CPU power. Following the results achieved in previous work [12, 13], in this experiment we have considered one single type of VM: large (m1.large – 7.5 GB RAM, 850 GB storage, 2 virtual cores). Each VM uses Linux Cent OS 5 (64-bit), and it was configured with the necessary software and bioinformatics libraries. All instances were configured to be accessed using Secure Shell (SSH) without password checking. The image ami-742bf91d is stored in the cloud and SciCumulus creates a virtual cluster to execute the experiment based on this image.

B. Experiment Setup

Our experiments use an input dataset of 132 multi-fasta files of protein-coding DNA sequences related to a set of 17 candidate target enzymes found in protozoan genomes. For more details about the enzymes, please refer to Ocaña *et al.* [13]. These sequences were previously obtained by hmmsearch comparisons. Then, hmmsearch comparisons retrieved homologous sequence hits that belong to the same 17 enzymes. This way, sequence hits were organized separately in files, grouped by enzyme and by species, summarizing a total of 132 multi-fasta file used in this experiment. All coding sequences were extracted from RefSeq (http://www.ncbi.nlm.nih.gov/RefSeq/) and Genbank (http://www.ncbi.nlm.nih.gov/Genbank/) databases. The following versions of the programs were used: MAFFT version 6.857, ReadSeq version 2.1.26, RAxML version 7.2.8-ALPHA and PAML version 4. All of them using default parameters. In our experiments we analyze the nested models (*i.e.* M0 *vs.* M3, M1 *vs.* M2, and M7 *vs.* M8) to identify which selective pressure is exerted in sequences, *i.e.*, if positive (Darwinian) pressure is (or not) present in data. We have inferred natural selection by estimating ω in SciEvol, considering $\omega < 1$, $\omega = 1$ and $\omega > 1$ representing purifying (negative) selection, neutral evolution and diversifying (positive) selection evolution, respectively [18]. The identification of genes whose ratio is greater than one is an evidence for adaptive evolution of the gene. The validity of this type of approach has been corroborated by recent reports of experimental verification using statistical predictions [19]. Whole genome data can facilitate the investigation of a particular trait or disease if genes are known *a priori*, or they can be used for global searches for evolution and adaptation of genes, which may reveal novel insights to species-specific biology. This way, by analyzing the selected input dataset, we expect to improve understanding the evolutionary history of this type of enzymes in protozoan genomes, defining their evolutionary selective pressure characteristics.

C. Biological Analysis: Results and Discussion

This sub-section presents biological results of the exploratory analysis using SciEvol. Most studies focusing on positive selection in pathogens have targeted specific genes that were candidates due to their functional relevance, such as those coding for antigenic proteins or genes involved in drug resistance [16]. More recently, studies have benefited from large sequence datasets to identify genes under positive selection without *a priori* candidates, particularly in pathogens [17, 24, 25]. Genes that show significant signals of positive selection were putatively involved *e.g.* in nutrient uptake from the host, secondary metabolite synthesis, respiration under stressful conditions, and regulation of expression by other genes. In our experiment, 35 of the 132 selected genes presented significant signal of positive selection, which represents a reasonable number of candidate genes for further investigation. In this paper we explain in detail just one of the enzymes used in this experiment, the 6-phosphogluconate dehydrogenase, decarboxylating (PGD) found in Trypanosomatids. The analysis of all other enzymes can be obtained at http://www.cos.ufrj.br/~kary. PGD is the third enzyme of the pentose phosphate biochemical pathway and it is also considered a potential drug target for Human African Trypanosomiasis. Expression of *T. brucei* PGD appears to be essential for the viability of this parasite [26]. Inhibition of the enzyme diminishes the cellular pool of NADPH (the reduced form of $NADP^+$, Nicotinamide adenine dinucleotide phosphate, used as a reducing agent) making the parasite more vulnerable to oxidative stress and increases the levels of 6-phosphogluconate.

To analyze PGD in Trypanosomatids, Table 1 presents the ML Estimates (MLEs) of parameters and the l values. Table 1 adopts log-likelihood l values and parameter estimated under models of variable ω ratios among sites. $2\Delta l$ statistical test is the difference between the l at the nested models. Seven was the number of sequences used. All information on Table 1 was obtained from the SciCumulus provenance repository. Model M0 assumes a single average ω across all sites and the estimate of $\omega = 0.06473$ indicated that purifying selection is the predominant evolutionary force acting on PGD. The LRT statistic value ($2\Delta l$) comparing models M0 *vs.* M3 is $2\Delta l = 180.82$, which is greater than critical values from x_4^2. MLEs under M2 suggest that sites are under neutral selection, however MLEs based on Bayes Empirical Bayes (BEB) probabilities also suggest that some sites are under positive selection.

As a refined test for selection, model M8 was compared with M7. Model M8 adds an additional site class of $\omega > 1$ to M7. M7 *vs.* M8 comparison provides another test of positive selection with $2\Delta l = 20.97$ much greater that critical values from the x_2^2. This comparison suggests the presence of several positively selected sites. The overall estimated ω value (1.26360) for the gene based on model M8 indicates positive selection. The emergence of selective MLEs along the gene (based on models M8 and M2) clearly indicates that the sites under positive selection are concentrated in six principal amino acid residues, i.e., 39E, 40S, 297S, 308F, 381Q, 405R. LRT for M7 *vs.* M8 indicates positive selection significant ($p < 0.0001$). Sites potentially under positive selection identified under model M8 and M2 are listed according to the PGD sequence in Table 1.

The NEB and BEB produced almost identical posterior probabilities and lists of positively selected sites. Furthermore, models M2 and M8 produced highly similar results (Table 1). The MLEs of parameters under the M8 model suggest that sites present $\omega_S = 1.264$ in proportion $p_l = 0.023$. These proportions are the prior probabilities that any site belongs to the classes. The data at a site (codon configurations in different sequences) alter the prior probabilities dramatically, so that the posterior probabilities can be very different from the prior probabilities. For example, at positive selected site 40S, the mean $\omega = 1.131$ and the mean posterior probability is 0.862, and thus the site is certainly to be under purifying selection. At

Table 1. PGD in Trypanosomatids

Model	l	$2\Delta l$	Estimates of parameters	Positively selected sites
M0-one-ratio	-5125.988694		$\omega = 0.06473$	Not allowed
M3-Discrete	-5035.576490	M0 *vs.* M3 180.824408 ($p < 0.0001^{\&\&}$)	$\omega_0 = 0.000$, $p_0 = 0.516$ $\omega_1 = 0.120$, $p_1 = 0.434$ $\omega_2 = 0.895$, $p_2 = 0.050$	None
M1a-Neutral	-5063.696841		$\omega_0 = 0.043$, $p_0 = 0.908$ $\omega_1 = 1.000$, $p_1 = 0.092$	Not allowed
M2a-Selection	-5063.696841	M1a *vs.* M2a 0 ($p = 1.0$)	$\omega_0 = 0.043$, $p_0 = 0.908$ $\omega_1 = 1.000$, $p_1 = 0.092$ $\omega_2 = 1.000$, $p_2 = 0.000$	39E, 40S, 297S, 300E, 301L, 308F, 314G, 381Q, 405R (by BEB probabilities)
M7-beta	-5039.003319		$\beta(0.271, 2.673)$	Not allowed
M8- β and ω	-5028.518936	M7 *vs.* M8 20.968766 ($p < 0.0001^{\&\&}$)	$\beta(0.323, 4.069)$ $p_0 = 0.977$, ($p_1 = 0.023$) $\omega_S = 1.264$	39E, 40S, 297S, 308F, 381Q, 405R (by NEB probabilities)

$^{\&\&}$Significant at $p < 0.05$ in plain text;

site 37K, the mean $\omega = 0.019$ and the mean posterior probability is 0.0 and thus the site is almost likely to be under strong purifying selection.

E. Performance Results

We have evaluated the performance of the parallel execution of SciEvol using SciCumulus workflow engine in a public cloud to show that: (i) the effective use of SciCumulus to execute SciEvol considerably improved the workflow performance; and (ii) clouds are suitable environments to execute parallel workflows that demand HPC, considering, of course, important limitations of the environment [9]. First we measured the performance of executing SciEvol on a single processor machine (one core) to analyze the local optimization before scaling up the number of VMs. Following, we have measured the performance and scalability of SciEvol by executing it using from 2 up to 64 large sizes (*m1.large*) VMs (thus totalizing 128 virtual cores). This experiment consumed as input 132 multi-fasta files in each execution. In Fig. 3 we present the measurements of total execution time (in hours) of SciEvol. The total workflow execution time decreases, in all cases, when SciCumulus provides more VMs (and consequently more virtual cores) to use. For example, when it processes 132 multi-fasta files of protein sequences using MAFFT, the total execution time was reduced from 232.85 hours (using one single virtual core) to 12.47 hours (using 128 virtual cores), which means performance improvements of up to 94.64%.

To evaluate the behavior of performance gain according to the number of processing units, we used speedup metric. An ideal speedup reduces the sequential time dividing this time by the number of processing units used. Speedup was defined to evaluate performance gains of parallel computers, in clouds there are many factors that can harm an ideal speedup. For example, in parallel computers the speedup value is impacted by serial portions of the code and communication between processors, while in the cloud, we have to consider other factors such as heterogeneity of the environment, performance fluctuations due to the virtualization and high communication latency [8]. The execution of SciEvol using 16 virtual cores led to a speedup of 5.76. Even though there was always a gain by adding more virtual cores, from 16 up to 128 cores, the speedup presented

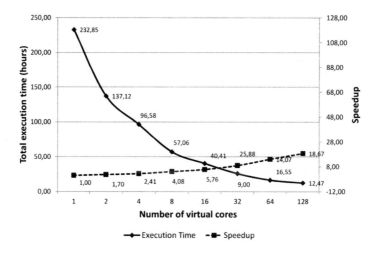

Fig. 3. Total execution time and speedup for SciEvol workflow

some degradation since the distribution of activity durations is heterogeneous. This result indicates that acquiring more VMs for execution may not bring the expected benefit, particularly if monetary costs are involved. We observed that, when the number of activities becomes closer to the number of VMs, many cores of the VMs may remain idle, thus not producing a positive impact on the total execution time.

5   Related Work

MER experiments modeled as scientific workflows are not yet fully explored in bioinformatics. Most of the existing approaches are based on individual scripts or Web applications. IDEA (Interactive Display for Evolutionary Analyses) [4] is one example. It is a graphical interface that edits and parses text input and output files. It interacts with the PHYLIP program [23] for phylogenetic maximum parsimony or neighbor-joining reconstruction trees and with codeml and baseml programs for MER analyses. Finally, IDEA can process data in parallel on a local machine or computing grid, allowing analyses to be completed quickly. However, IDEA is not capable of executing experiments in terms of workflows and it does not concern about extracting and managing provenance data. Another example is PhyleasProg [5], a user-friendly Web server dedicated to evolutionary analyses. Scientists need to enter a list of Ensembl database protein IDs and a list of species as inputs. MSA are performed by Muscle and refined by GBlocks program [27]. Phylogenetic trees are reconstructed by TreeBeST program [28] and codeml performs positive selection computation. Compared with these approaches, SciEvol presents the advantage of structuring the experiment in a parallel workflow with provenance associated. Provenance information is available for scientists that use SciEvol to query, without requiring any additional effort. All of this is supported by a provenance database that stores fundamental information about the workflow trials.

6   Final Remarks

MER workflows are required in the bioinformatics field to explore genomes and to infer the evolutionary history of a pre-determined set of genes and enzymes of interest. Some of the activities within these MER analysis workflows may execute for several weeks, thus requiring HPC environments and parallel techniques. In this paper, we presented SciEvol workflow that is focused on MER analysis with thorough experimental evaluation from biological and computational perspectives. SciEvol aims at estimating parameters and testing hypotheses for studying evolutionary process. Since SciEvol activities are computing and data-intensive, it was designed and implemented in SciCumulus, a cloud-based parallel workflow engine, to be executed in parallel on clouds.

In order to evaluate SciEvol we performed biological and computational studies. From the biological perspective, our evaluation showed that 35 of the 132 genes found in protozoan genomes are under positive selection. There has been particular interest in identifying positive selection. While negative selection is pervasive in functional genetic elements, positive selection provides evidence for adaptive changes in function. Genes under positive selection in pathogens have been a priority in efforts to investigate coevolution dynamics and to develop vaccines or drugs [16, 17, 24, 25]. This way, genes selected in this study provide hypotheses that could be used as primary targets of positive selection refinement and structural studies due to their role in interactions with the host immune/defense system. It can be subsequently interpreted in the context of protein structure and function, cellular localization or other appropriate attributes of the gene in question.

The other analysis focuses on the computational perspective. In this evaluation, we have executed SciEvol processing 132 multi-fasta files in each execution. Thus, this paper also contributes by showing the potential of executing computing and data-intensive scientific workflows in parallel at environments such as clouds. These executions involve several hundreds of tasks and data files that produced several gigabytes of data. Using provenance data from SciCumulus engine, it was possible to query information from this large volume of tasks and data files. For example, scientists may query whether the results obtained using model M8 present probability $P > 95\%$ or $P > 99\%$ of coming from the site class of positive selection. Depending on the results, scientists decide if they have to re-execute SciEvol, exploring other parameters or to stop current execution. By analyzing the overall performance we can state that parallel techniques improved the performance of the workflow up to 94.64% when compared to a sequential workflow execution. The analysis of speedup showed that adding VMs always reduced total execution time. However, the cost/benefit of adding VMs needs to consider several aspects, such as, the number of activities to execute, the monetary cost and expected gains. We believe that workflows, in different bioinformatics areas such as structural, comparative and functional genomics, which require the exploration of large amounts of data in the experiments, can take advantage of the same resources SciEvol has explored in SciCumulus engine.

Acknowledgements. The work was partially funded by the Brazilian agencies CAPES, FAPERJ and CNPq.

References

1. Miller, W., Makova, K.D., Nekrutenko, A., Hardison, R.C.: Comparative Genomics. Annu. Rev. Genom. Human Genet. 5, 15–56 (2004)
2. Taylor, I.J., Deelman, E., Gannon, D.B., Shields, M.: Workflows for e-Science: Scientific Workflows for Grids. Springer (2007)
3. Freire, J., Koop, D., Santos, E., Silva, C.T.: Provenance for Computational Tasks: A Survey. Computing in Science and Engineering 10, 11–21 (2008)
4. Egan, A., Mahurkar, A., Crabtree, J., Badger, J.H., Carlton, J.M., Silva, J.C.: IDEA: Interactive Display for Evolutionary Analyses. BMC Bioinformatics 9, 524 (2008)
5. Busset, J., Cabau, C., Meslin, C., Pascal, G.: PhyleasProg: a user-oriented web server for wide evolutionary analyses. Nucleic Acids Research 39, W479–W485 (2011)
6. Katoh, K., Toh, H.: Recent developments in the MAFFT multiple sequence alignment program. Brief. Bioinformatics 9, 286–298 (2008)
7. Goldman, N., Yang, Z.: A codon-based model of nucleotide substitution for protein-coding DNA sequences. Mol. Biol. Evol. 11, 725–736 (1994)
8. Hey, T., Tansley, S., Tolle, K.: The Fourth Paradigm: Data-Intensive Scientific Discovery. Microsoft Research (2009)
9. Vaquero, L.M., Rodero-Merino, L., Caceres, J., Lindner, M.: A break in the clouds: towards a cloud definition. SIGCOMM Comput. Commun. Rev. 39, 50–55 (2009)
10. Jackson, K.R., Ramakrishnan, L., Runge, K.J., Thomas, R.C.: Seeking supernovae in the clouds: a performance study. In: Proceedings of the 19th ACM International Symposium on High Performance Distributed Computing, pp. 421–429. ACM, New York (2010)
11. Yang, Z.: PAML 4: phylogenetic analysis by maximum likelihood. Mol. Biol. Evol. 24, 1586–1591 (2007)
12. Ocaña, K.A.C.S., de Oliveira, D., Dias, J., Ogasawara, E., Mattoso, M.: Optimizing Phylogenetic Analysis Using SciHmm Cloud-based Scientific Workflow. In: 2011 IEEE Seventh International Conference on e-Science (e-Science), pp. 190–197. IEEE, Stockholm (2011)
13. Ocaña, K.A.C.S., de Oliveira, D., Ogasawara, E., Dávila, A.M.R., Lima, A.A.B., Mattoso, M.: SciPhy: A Cloud-Based Workflow for Phylogenetic Analysis of Drug Targets in Protozoan Genomes. In: Norberto de Souza, O., Telles, G.P., Palakal, M. (eds.) BSB 2011. LNCS (LNBI), vol. 6832, pp. 66–70. Springer, Heidelberg (2011)
14. de Oliveira, D., Ogasawara, E., Baião, F., Mattoso, M.: SciCumulus: A Lightweight Cloud Middleware to Explore Many Task Computing Paradigm in Scientific Workflows. In: 3rd International Conference on Cloud Computing, pp. 378–385. IEEE Computer Society, Washington, DC (2010)
15. Anisimova, M., Bielawski, J.P., Yang, Z.: Accuracy and power of the likelihood ratio test in detecting adaptive molecular evolution. Mol. Biol. Evol. 18, 1585–1592 (2001)
16. Aguileta, G., Refrégier, G., Yockteng, R., Fournier, E., Giraud, T.: Rapidly evolving genes in pathogens: methods for detecting positive selection and examples among fungi, bacteria, viruses and protists. Infect. Genet. Evol. 9, 656–670 (2009)

17. King, C.-C., Chao, D.-Y., Chien, L.-J., Chang, G.-J.J., Lin, T.-H., Wu, Y.-C., Huang, J.-H.: Comparative analysis of full genomic sequences among different genotypes of dengue virus type 3. Virol. J. 5, 63 (2008)
18. Nielsen, R., Yang, Z.: Likelihood models for detecting positively selected amino acid sites and applications to the HIV-1 envelope gene. Genetics 148, 929–936 (1998)
19. Yang, Z.: Computational Molecular Evolution. Oxford University Press (2006)
20. Freedman, D., Pisani, R., Purves, R.: Statistics, 4th edn. W. W. Norton (2007)
21. Muse, S.V., Gaut, B.S.: A likelihood approach for comparing synonymous and nonsynonymous nucleotide substitution rates, with application to the chloroplast genome. Mol. Biol. Evol. 11, 715–724 (1994)
22. Yang, Z., Swanson, W.J.: Codon-substitution models to detect adaptive evolution that account for heterogeneous selective pressures among site classes. Mol. Biol. Evol. 19, 49–57 (2002)
23. Felsenstein, J.: PHYLIP - Phylogeny Inference Package (Version 3.2). Cladistics 5, 164–166 (1989)
24. Chen, S.L., Hung, C.-S., Xu, J., Reigstad, C.S., Magrini, V., Sabo, A., Blasiar, D., Bieri, T., Meyer, R.R., Ozersky, P., Armstrong, J.R., Fulton, R.S., Latreille, J.P., Spieth, J., Hooton, T.M., Mardis, E.R., Hultgren, S.J., Gordon, J.I.: Identification of genes subject to positive selection in uropathogenic strains of *Escherichia coli*: a comparative genomics approach. Proc. Natl. Acad. Sci. U.S.A. 103, 5977–5982 (2006)
25. Ge, G., Cowen, L., Feng, X., Widmer, G.: Protein coding gene nucleotide substitution pattern in the apicomplexan protozoa *Cryptosporidium parvum and Cryptosporidium hominis*. Comp. Funct. Genomics 879023 (2008)
26. Montin, K., Cervellati, C., Dallocchio, F., Hanau, S.: Thermodynamic characterization of substrate and inhibitor binding to *Trypanosoma brucei* 6-phosphogluconate dehydrogenase. FEBS J. 274, 6426–6435 (2007)
27. Talavera, G., Castresana, J.: Improvement of phylogenies after removing divergent and ambiguously aligned blocks from protein sequence alignments. Syst. Biol. 56, 564–577 (2007)
28. Vilella, A.J., Severin, J., Ureta-Vidal, A., Heng, L., Durbin, R., Birney, E.: EnsemblCompara GeneTrees: Complete, duplication-aware phylogenetic trees in vertebrates. Genome Res. 19, 327–335 (2009)

How Bioinformatics Enables Livestock Applied Sciences in the Genomic Era

José Fernando Garcia[1,*], Adriana Santana do Carmo[1,2], Yuri Tani Utsunomiya[1,2], Haroldo Henrique de Rezende Neves[1,3], Roberto Carvalheiro[4], Curtis Van Tassell[5], Tad Stewart Sonstegard[5], and Marcos Vinicius Gualberto Barbosa da Silva[6]

[1] UNESP – Univ. Estadual Paulista – Departamento de Apoio, Produção e Saúde Animal – FMVA – Campus Araçatuba – SP – Brazil
[2] UNESP – Univ. Estadual Paulista – Departamento de Reprodução Animal – FCAV – Campus Jaboticabal – SP – Brazil
[3] UNESP – Univ. Estadual Paulista – Departamento de Zootecnia – FCAV – Campus Jaboticabal – SP – Brazil
[4] GenSys Associated Consultants. Porto Alegre – RS – Brazil
[5] USDA-ARS, ANRI, Bovine Functional Genomics Laboratory, Beltsville, Maryland 20705, USA
[6] Embrapa Dairy Cattle - Juiz de Fora - MG - Brazil

Abstract. This review paper presents the three main approaches currently used in livestock genomic sciences where the bioinfomatics plays a critical role. They are named as Genomic Selection (GS), Genome Wide Association Study (GWAS) and Signatures of Selection (SS). The subsides for the construction of this article were generated in a current project (started in 2011), so called Zebu Genome Consortium (ZGC), which joins researchers from different institutions and countries, aiming to scientifically explore genomic information of *Bos taurus indicus* cattle breeds and deliver useful information to breeders and academic community, specially from the tropical regions of the world.

Keywords: Genomic Selection, Genome Wide Association Study, Signatures of Selection, Livestock, Tropical Environment, *Bos taurus indicus*.

1   Introduction

There has been much progress in genomic sciences since a draft sequence of the human genome was published just ten years ago. Due to this effort for decoding the human genome, opportunities for understanding the basic biology of animal health and production are now unparalleled for some species, as advances in genomics are exploited to obtain comprehensive foundational knowledge about the structure and function of these genomes, and about the genetic contributions to phenotypes

---

* Corresponding author.

M.C.P. de Souto and M.G. Kann (Eds.): BSB 2012, LNBI 7409, pp. 192–201, 2012.

underlying production, health and disease. Recent projections and modeling of population growth and food demands have suggested agribusiness must double food production in the next 40 years to successful sustain the human population and avoid catastrophic overpopulation. Some experts predict that increases in animal protein production will have to make up about 70% of this doubling in food production due to increasing demand from consumers with more disposable income and limitations on the amount of arable land available for future crop production [1]. Thus, it is somewhat surprising at this time of the genomics revolution that opportunities afforded by advances in technology and new concepts for genetic improvement based on unique population structure in domesticated animals have been overlooked. This is especially evident when considering the return on investment already generated for DNA assisted genetic improvement from very minimal funding. More importantly, there is a large gap in expertise due to an overall lack of animal scientists that possess the knowledge and skills to apply modern genome analysis methods to problems for improving animal efficiency and product quality.

Based on these facts and observations, we believe the large-scale study of livestock genetic resources enabled by computational biology can provide the innovation needed to improve animal food production. We present three main whole genome scan approaches the Zebu Genome Consortium has been applying to the Brazilian Nellore cattle (*Bos taurus indicus*) in order to achieve this innovation: Genomic Selection (GS), Genome Wide Association Study (GWAS) and Signatures of Selection (SS).

2   Genomic Selection: Increasing Accuracy of Genetic Predictions in Cattle

The possibility of making accurate predictions of the genetic merit of individuals by using genotypes based on dense single nucleotide polymorphism (SNP) marker panels, a process known as genomic selection (GS), is nowadays revolutionizing the design and implementation of livestock breeding programs. Conceived by [2], who provided the theoretical basis for this approach even before the existence on high density SNP panels, and further boosted as described by [3], who highlighted potential benefits of this strategy in reducing generation intervals, increasing both accuracies of prediction and selection intensities, reducing costs of breeding organizations and making feasible genetic evaluation for difficult-to-measure traits.

The logic behind genomic selection approach is that if the SNP marker density is high enough to cover the entire genome, most of the quantitative trait loci (QTL) will be in linkage disequilibrium with some markers. Therefore, the sum of all marker effects (direct genomic value, DGV) will be a good predictor of a given animal genetic merit and should enable selection decisions as soon as genotypes of such animals are available (just after birth or even during the embryo/fetus developmental phases).

Following these expectations, the sequencing of bovine genome [4] and the availability of dense SNP marker panels [5], allowed GS to migrate from simulation to real-word. First successful applications of GS were verified in North American dairy cattle [6-7] and motivated studies on GS in other breeds and populations. As the SNP panel density increases, the amount of information to be processed poses new challenges from statistical and computational points of view, especially due to the number of predictor variables (markers) which are generally much higher than the number of observations (phenotypes), leading to the lack of degrees of freedom to estimate all marker effects simultaneously and also giving rise to multi co-linearity problems.

Although there is extensive list of scientific papers comparing different statistical methods applied to genomic selection in dairy cattle [8-10], there is still a lack of such comparisons in beef breed panels to define the best method for genetic predictions. In addition, most of the GS studies in cattle were carried out in *Bos taurus taurus* breeds (those mostly common in temperate climate regions of the world), which would not apply directly to *Bos taurus indicus* breeds (those mainly used in tropical regions of the world – Latin America, Africa and Asia).

Specifically in Brazil, the beef breed Nellore plays an important role in meat production under tropical systems and it is expected that genome-enabled predictions (GS) could contribute considerably to improve the efficiency of production in such systems. This breed has a large number of recorded animals and, although has experienced significant genetic progress for growth traits in the last two decades, the progress achieved by means of conventional selection is still low for traits related to reproduction, meat quality and feed efficiency. Due to these particularities, the Zebu Genome Consortium is testing the feasibility of applying GS in Nellore cattle breeding programs, by comparing accuracies and prediction biases observed using four different statistical methods for 17 traits of economic relevance in this breed.

As an illustration of the power of GS approach, Figure 1 shows preliminary results from ZGC pointing out to the changes in accuracies observed when using genomic information for genetic prediction estimation in Brazilian Nellore cattle. Accuracies observed in genomic predictions are reaching the expectations (0.5 in average, with some traits showing 0.7), seeming to be high enough to motivate further action towards the incorporation of this technology in breeding schemes aiming to enhance selection decisions in this breed.

As genomic predictions of genetic values are based on computationally expensive procedures, fine programming for efficient calculations and management of hardware resources has been a constant pursuit. Additionally, the performance of methods for the estimation of marker effects may vary depending on the different genetic architecture of the trait, which implies no optimum general analytical procedure. This demands testing multiple methods for each single trait, bringing exponentiation of computation constraints. In this context, Bioinformatics knowledge is an essential tool for GS studies.

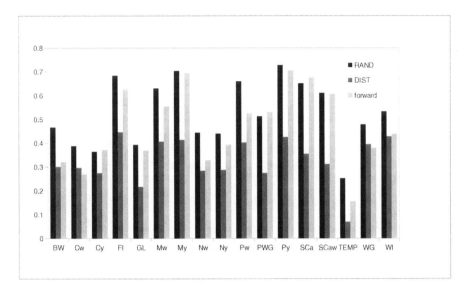

Fig. 1. Example of empirical accuracies[1] of genomic predictions in Nellore cattle for 17 traits[2] as a function of validation strategy[3] (Zebu Genome Consortium communication). [1]Empirical accuracies calculated as the Pearson's correlation between DGV and EBV (estimated breeding value) of year 2011 for bulls in the testing set. In the case of cross-validation strategies the pooled average of empirical accuracies over the 5-folds was plotted. [2]WG: weight gain from birth to weaning (age about 205 days); Cw, Pw, Mw, Nw: visual scores taken at weaning for carcass conformation, finishing precocity, muscling and navel, respectively; PWG: weight gain from weaning to long-yearling (age about 550 days); Cy, Py, My, Ny: visual scores taken at long-yearling for carcass conformation, finishing precocity, muscling and navel, respectively; TEMP: score for temperament; SCa and SCaw: scrotal circumference adjusted for age and for age and weight, respectively. BW: birth weight; GL: gestation length; WI: weaning index, composed by traits evaluated at weaning; FI: final index, composed by traits evaluated at weaning and long-yearling (FI). [3]RAND: 5-fold cross-validation, splitting animals randomly in groups of similar size. DIST: 5-fold cross-validation, based on k-means clustering of animals based on their genomic distance (i.e. minimizing inter-groups relationships). Forward: training set composed by bulls with high accuracy in 2007 and the remainder bulls (with high accuracy in 2011) composed the testing set.

3   Genome Wide Association Study for Disease Resistance and Complex Traits in Cattle

Genome Wide Association Study (GWAS) can be defined as the use of dense SNP marker panels to survey a given population for the genetic variations that could play a role in virtually any phenotypic characteristic [11]. GWAS using the current high density SNP marker platforms tend to be cheaper, simpler and more effective than the methods usually used in the past, which relied in the candidate gene approach, and has been proposed as a powerful mean to identify the common variants that underlie complex traits [12]. In general, this approach represents new challenges in data interpretation, terminology and statistical models, requiring essential elements to be successfully implemented [13].

One of these elements is the need to sample populations in which the phenotype of interest is representative. Sample size should be large enough to detect even small associations with the SNP variants, considering that some polymorphism that eventually explains only 1% of the phenotype can be effective in the elucidation of phenotype's biological basis [14].

Another critical element in GWAS is the use of statistical methods powerful enough to detect associations without major biases. Also, the computational requirements are high in this type of analysis and the use of combined bioinformatics resources is very important to make an association study feasible [15].

The proper use of the statistical methods and genetic assumptions during the analysis allow the identification of genomic regions strongly related to a given phenotype, and pointing out genes involved in the analyzed characteristic (Figure 2).

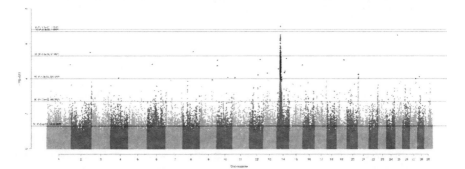

Fig. 2. Example of a "Manhattan plot" generated from a GWAS study involving a Nellore cattle population analyzed with a high density SNP marker panel (~700.000 SNP). Dots represent individual SNP horizontally distributed in the chromosome order (from 1 to 29). Gray (dark and light) represents SNP that did not pass the statistical significance threshold chosen in the given study. Blue dots represent SNP x phenotype significant associations and horizontal lines the different levels of significance commonly adopted in GWAS studies. A strong association can be observed in chromosome 14 (Zebu Genome Consortium communication).

However, challenges regarding GWAS approach are beyond the processing and analyzing of data and generating "Manhattan plots", but especially related to those involving its interpretation. In this context, new techniques for data mining, able to extract consistent biological information from raw data, are needed to interpret the biological significance of a given GWAS finding. The knowledge generated by GWAS can later be used in areas such as transcriptomics, epigenomics, proteomics and metabolomics. The dynamic combination of information from these studies allows for integrating the different parts of the whole, creating a concept called Systems Biology [16].

There are several tools available to assist in the annotation of genes underlying associations. Such tools provide information on the existing bibliography, expression patterns and information about homologous and interacting genes. When using bioinformatics approaches to assess biological candidacy, a researcher will rely on extracting as much accurate information possible from online database sources [17].

Essential aspect of these approaches is the translation of chromosomal coordinates and candidate genes, into of information that may aid the understanding of biological mechanisms. In recent years, several methods and tools have been developed to interpret long lists of genes or proteins using information available in biological databases.

The need for a biological interpretation of GWAS results created a new approach called Functional Enrichment Analysis. Basically, this approach aims to evaluate the frequency of functional terms from the list of genes, applying statistical tests to determine those significantly overrepresented or functionally enriched [18]. Cured functional information from different sources, such as Gene Ontology and KEGG databases are commonly used in that context [19,20]. These sources have maps and information that bring together common patterns of interactions between different system components. Several tools were developed for the application of functional enrichment analysis, although each application could introduce their own sources of variation, such as different statistical tests, terms and sources of information available for different organisms, in general they have the same final results [18].

4 Signatures of Selection and Their Role in Unraveling Traits under Selection in Livestock

Natural selection implies that beneficial traits spread out quickly through a population, because animals that are more fitted to the environment have higher probability of survival and thus leave more offspring. Considering animals living under the same constant environment, trait variation may occur due to genetic differences among individuals. A genome selective sweep happens either when the environment changes and a pre-existent neutral DNA variant turns into advantageous in the new condition or when a beneficial mutation arises and is passed to the future generations. Both processes force patterns of allele frequency in the surrounding DNA sequence variation, which are called signatures of selection (SS) (also referred as footprints of selection). This kind of signal differs from what is expected from the background neutral DNA sequence, as recombination and random mating tend to shuffle the neutral variants within the genome. Large-scale genotype and whole genome sequence (WGS) data can help detecting such patterns and reveal genomic regions harboring variants under selection.

SS emerge from basically three different selection phenomena [21]: (1) positive selection (a single allele is advantageous over others), (2) balancing selection (multiple alleles exhibit advantage together, e.g. heterozygote advantage) and (3) purifying selection (elimination of a disadvantageous allele). The outcome pattern depends on the underlying selection phenomenon, the pressure intensity and the age of the selection process. Although signals can be identified within a population, some cases require multiple population comparisons [22-23] or even multiple methods combination [24] to detect the footprint. Table 1 summarizes the main SS detectable patterns and cites some methodologies suitable for their identification from genomic data.

Analytical algorithms have been developed for genome-wide scan of SS, and many of those have been implemented in open-source stand-alone software, scripts or packages by the authors proposing them. However, these tools are frequently dedicated to users with moderate to advanced knowledge in both programming and biology, as they not only require handling and preparation of large-scale input data, but also demand expertise on efficiently parsing meaningful results and sometimes modification of the source code for better performance. Some methods do not have computational tools available, and bioinformatics expertise is once more demanded for development of homegrown code.

In livestock breeding programs, animals are being artificially selected for traits of human interest under intense pressure. The dissemination of the selected variants is quicker in livestock when compared with human or wild life populations due to assisted reproduction, which may accelerate the formation of genomic footprints. Thus, genome-wide scans for SS can be an alternative approach to GWAS to identify which portions of the DNA are sheltering genes that are being recently selected. The identified region can be further linked together with information available in the literature and public databases, allowing for functional annotation of the underlying molecular features involved with the selection phenomenon. Table 2 lists the results of five interesting studies from genome-wide scan of SS in livestock species.

Table 1. Signatures of Selection types detectable from genomic data. PS. Positive Selection, PU. Purifying Selection, BS. Balancing Selection.

Type of signature	Detectable pattern	Methodologies	Underlying selection phenomena	Age of selection (generations)
Function-altering mutation	changes in non-synonymous to synonymous variation ratio in the open reading frame of a coding region.	$\omega = D_N/D_S$ (Nielsen & Yang, 1998)	PS and PU	> 40,000
Local genomic diversity loss	deficit of local heterozygosity compared to the rest of the genome within and between populations	ZH_p (Rubin et al., 2010)	PS	> 10,000
Change in the allele frequency spectrum	increase in the frequency of derived alleles within and between populations	ΔDAF (Grossman et al., 2010), *Tajima's D* (Tajima, 1989), *CLR* (Williamson et al., 2007)	PS	< 3,200
Population differentiation	difference in the allele frequency between populations	F_{ST} (Weir & Cockerham, 1984)	PS and BS	< 3,000
Extended linkage disequilibrium	linkage disequilibrium persistancy and long-range haplotypes within and between populations	*LRH* (Sabeti et al., 2002), *iHS* (Voight et al., 2006), *XP-EHH* (Sabeti et al., 2007), *Rsb* (Tang et al., 2007), *ΔiHH* (Grossman et al., 2010), *varLD* (Ong & Teo, 2010)	PS	< 1,200

Table 2. Studies applying genome-wide scan for Signatures of Selection identification in livestock species

Candidate genes	Function and hypothesized selection pressure	Reference	Type of genomic data	Methods applied	Species and breeds
TSHR	Regulation of photoperiod control of reproduction. Unseasonal, light-independent reproduction	Rubin et al. 2010	WGS, 5x coverage	ZH_p	Domestic lines of chicken
RXFP2	Sexual maturation and testicular descent. Reproduction under warm conditions	Gautier & Naves 2011	44k SNPs	Rsb and local ancestry	Creole cattle from Guadalupe
ACTC1, COL23A1, MATN2, and FAP	Muscle formation. Beef production	Qanbari et al. 2011	50k SNPs	iHS and F_{ST}	Sets of dairy and beef cattle breeds
Clusters of genes related to the somatotropic and the gonadotropic systems	Milk metabolism and reproduction. Milk production and antagonic effect on fertility	Flori et al. 2009	42k SNPs	F_{ST}	Dairy cattle breeds
RXFP2	Sexual maturation and bone mass. Horn absence	Kijas et al. 2012	49k SNPs	F_{ST}	74 sheep breeds

5   Concluding Remarks

Among the livestock species, bovine is the one with more achievements coming from the application of genomics in its routine processes. Bioinformatics plays a key role in the continuous development of animal genomics, and the association of information generated from Genomic Selection, Genome Wide Association Study and Signatures of Selection approaches should allow the integration of different data sources and the better understanding of the mechanisms controlling phenotype manifestation. With these integrated information, it would be possible to refine the processes used to select and breed animals, and consequently help in the improvement of lives through the increase of animal protein to feed the growing world population.

Acknowledgements. Besides the present article authors, Zebu Genome Consortium is formed by: John Cole (USDA – ARS – USA), John McEwan (AgResearch New Zealand), Johannes Soelkner / Ana Maria Perez O'Brien (BOKU – Austria) and Flávio Schenkel (UofGuelph – Canada), to whom the authors acknowledge for the constant scientific inspiration and contribution. CNPq (578738/2008-2, 475914/2010-4, 560922/2010-8, 483590/2010-0), FAPEMIG (CVZ PPM 0079/2010, 12093/2010) and FAPESP (2010/52030-2, 2010/51975-3) should be acknowledged for the financial support to our work.

References

1. FAO (Food and Agriculture Organization of United Nations) (2012),
   http://www.fao.org/fileadmin/templates/wsfs/docs/expert_paper/
   How_to_Feed_the_World_in_2050.pdf
2. Meuwissen, T.H.E., Hayes, B.J., Goddard, M.E.: Prediction of total genetic value using genome-wide dense marker maps. Genetics 157, 1819–1829 (2001)
3. Schaeffer, L.R.: Strategy for applying genome-wide selection in dairy cattle. J. Anim. Breed. Genet. 123, 218–223 (2006)
4. Bovine Genome Sequencing and Analysis Consortium. The genome sequence of taurine cattle: a window to ruminant biology and evolution. Science 324(5926), 522–528 (2009)
5. Van Tassell, C.P., Smith, T.P., Matukumalli, L.K., Taylor, J.F., Schnabel, R.D., Lawley, C.T., Haudenschild, C.D., Moore, S.S., Warren, W.C., Sonstegard, T.S.: SNP discovery and allele frequency estimation by deep sequencing of reduced representation libraries. Nature Methods 5(3), 247–252 (2008)
6. Harris, B.L., Johnson, D.L., Spelman, R.J.: Genomic selection in New Zealand and the implications for national genetic evaluation. In: Proc. Interbull Meeting, Niagara Falls, Canada (2008)
7. VanRaden, P.M., Van Tassell, C.P., Wiggans, G.R., Sonstegard, T.S., Schnabel, R.D., Taylor, J.F., Schenkel, F.S.: Invited review: Reliability of genomic predictions for North American Holstein bulls. J. Dairy Sci. 92, 16–24 (2009)
8. Moser, G., Tier, B., Crump, R.E., Khatkar, M.S., Raadsma, H.W.: A comparison of five methods to predict genomic breeding values of dairy bulls from genome-wide SNP markers. Genet. Sel. Evol. 41, 56 (2009)
9. Luan, T., Woolliams, J.A., Lien, S., Kent, M., Svendsen, M., Meuwissen, T.H.: The accuracy of genomic selection in Norwegian Red cattle assessed by cross-validation. Genetics 183, 1119–1126 (2009)
10. Legarra, A., Robert-Granié, C., Croiseau, P., Guillaume, F., Fritz, S.: Improved Lasso for genomic selection. Genet. Res. 93(1), 77–87 (2011)
11. Frazer, K.A., et al.: Human genetic variation and its contribution to complex traits. Nature Reviews 10, 241–251 (2009)
12. Hirschhorn, J.N., Daly, M.J.: Genome-wide association studies for common diseases and complex traits. Nature Reviews 6(1), 95–198 (2005)
13. Pearson, T.A., Manolio, T.A.: How to Interpret a Genome-wide Association Study. Journal of American Medical Association 299(11), 1335–1344 (2008)
14. Cantor, R.M., et al.: Prioritizing GWAS Results: A Review of Statistical Methods and Recommendations for Their Application. The American Journal of Human Genetics 86, 6–22 (2010)
15. Chan, E.K.F., et al.: The combined effect of SNP-marker and phenotype attributes in genome-wide association studies. Animal Genetics 40, 149–156 (2009)
16. Chuang, H., et al.: A decade of Systems Biology. Reviews in Advance 20(1), 15–20 (2010)
17. Webber, C.: Functional Enrichment Analysis with Structural Variants: Pitfalls and Strategies. Cytogenet. Genome Res. 135, 277–285 (2011)
18. Fontanillo, C., et al.: Functional Analysis beyond Enrichment: Non-Redundant Reciprocal Linkage of Genes and Biological Terms. PloS ONE 6(9), 242 (2011)
19. The gene ontology consortium. Gene ontology: tool for the unification of biology. Nature Genetics 25(1), 25–29 (2000)

20. Kanehisa, M., Goto, S.: KEGG: Kyoto Encyclopedia of Genes and Genomes. Nucleic Acids Res. 28(1), 27–30 (2000)
21. Oleksyk, T.K., Smith, M.W., O'Brien, S.J.: Genome-wide scans for footprints of natural selection. Phil. Trans. R. Soc. B 365, 185–205 (2010)
22. Sabeti, P.C., et al.: Genome-wide detection and characterization of positive selection in human populations. Nature 449, 913–918 (2007)
23. Tang, K., Thornton, K.R., Stoneking, M.: A new approach for using genome scans to detect recent positive selection in the human genome. PLoS Biology 5, e171 (2007)
24. Grossman, S.R., et al.: A composite of multiple signals distinguishes causal variants in regions of positive selection. Science 327, 883–886 (2010)
25. Nielsen, R., Yang, Z.: Likelihood models for detecting positively selected amino acid sites and applications to the HIV-1 envelope gene. Genetics 148, 929–936 (1998)
26. Rubin, C.J., et al.: Whole-genome resequencing reveals loci under selection during chicken domestication. Nature 464, 587–593 (2010)
27. Tajima, F.: Statistical method for testing the neutral mutation hypothesis by DNA polymorphism. Genetics 123, 585–595 (1989)
28. Williamson, S.H., Hubisz, M.J., Clark, A.G., Payseur, B.A., Bustamante, C.D., Nielsen, R.: Localizing recent adaptive evolution in the human genome. PLoS Genet. 3, e90 (2007)
29. Weir, B.S., Cockerham, C.C.: Estimating F-Statistics for the analysis of population structure. Evolution 38(6), 1358–1370 (1984)
30. Sabeti, P.C., et al.: Detecting recent positive selection in the human genome from haplotype structure. Nature 419, 832–837 (2002)
31. Voight, B.F., Kudaravalli, S., Wen, X., Pritchard, J.K.: A map of recent positive selection in the human genome. PLoS Biol. 4, e72 (2006)
32. Ong, R.T.-H., Teo, Y.Y.: varLD: a program for quantifying variation in linkage disequilibrium patterns between populations. Bioinformatics 26(9), 1269–1270 (2010)
33. Gautier, M., Naves, M.: Footprints of selection in the ancestral admixture of a New World Creole cattle breed. Molecular Ecology 20, 3128–3143 (2011)
34. Qanbari, S., et al.: Application of site and haplotype-frequency based approaches for detecting selection signatures in cattle. BMC Genomics 12, 318 (2011)
35. Flori, L., et al.: The genome response to artificial selection: a case study in dairy cattle. PLoS ONE 4(8), e6595 (2009)
36. Kijas, J.W., et al.: Genome-wide analysis of the world's sheep breeds reveals high levels of historic mixture and strong recent selection. PLoS Biology 10(2), e1001258 (2012)

Author Index